THE AIRLINE TRAINING PILOT

Nav2001
AERONAUTICAL INFORMATION NAVDATA DATABASE AND CHARTS

4. WAYPOINTS (Cont)
WAYPOINT DATABASE IDENTIFIERS (Cont)

- Approach Charts

    VNAV descent angle information derived from the Jeppesen NavData database is being added to approach charts. Identifiers are shown for the Final Approach Fix (FAF), Missed Approach Point (MAP), and the missed approach termination point.

    State-named Computer Navigation Fixes (CNFs) are shown on all applicable charts.

    GPS (GNSS) type approach charts include all database identifiers.

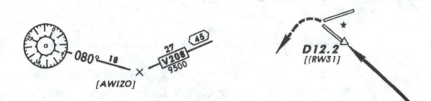

COMMON WAYPOINT NAME FOR A SINGLE LOCATION
Government authorities may give a name to a waypoint at a given location, but not use the name at the same location on other procedures in the same area. The Jeppesen NavData database uses the same name for all multiple procedure applications. Charting is limited to the procedure/s where the name is used by the authorities.

FLY-OVER versus FLY-BY FIXES/WAYPOINTS
In most cases, pilots should anticipate and lead a turn to the next leg. The database indicates when the fix must be crossed (flown-over) before the turn is commenced. The fix is coded as fly-over when the requirement is inferred or is specified by the governing authority. Fixes are charted as fly-over fixes only when specified by the governing authority.

Fly-over fixes have a circle around the fix/waypoint symbol. No special charting is used for fly-by fixes.

ULOGO and ROTGO
  Are fly-by waypoints.

RW03 and LESOV
  Are fly-over waypoints.

Nav2001
AERONAUTICAL INFORMATION NAVDATA DATABASE AND CHARTS

5. AIRWAYS

ATS ROUTES

Airways identified as ATC routes by States (countries) cannot be uniquely identified. They are not included in the Jeppesen NavData database.

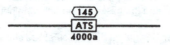

DESIGNATORS

Jeppesen NavData database airway designators are followed by a code indicating ATC services (such as A for Advisory, F for Flight Information) when such a code is specified by the State (country). Not all airborne systems display the ATC services suffix.

ALTITUDES

Minimum Enroute Altitudes (MEAs), Minimum Obstacle Clearance Altitudes (MOCAs), Off Route Obstacle Clearance Altitudes (OROCAs), Maximum Authorized Altitudes (MAAs), Minimum Crossing Altitudes (MCAs), Minimum Reception Altitudes (MRAs), and Route Minimum Route Off-Route Altitudes (Route MORAs) - - These minimum altitudes for airways are not displayed in most avionics systems.

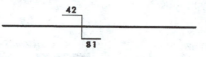

CHANGEOVER POINTS

Changeover points (other than mid-point between navaids) are on charts but are not included in the Jeppesen NavData database.

Nav2001
AERONAUTICAL INFORMATION NAVDATA DATABASE AND CHARTS

6. ARRIVALS AND DEPARTURES

PROCEDURES NOT IN THE DATABASE

Jeppesen publishes some officially designated departure procedures that include only text on IFR airport charts beneath the take-off minimums. They may be labeled "Departure Procedure", "IFR Departure Procedure", or "Obstacle DP". Most of these are U.S. and Canadian procedures, although there is a scattering of them throughout the world. Any waypoint/fix mentioned in the text is in the Jeppesen NavData database. *However, these text-only departure procedures are not in the database.*

TAKE-OFF & OBSTACLE DEPARTURE PROCEDURE			
	Rwy 17		**Rwy 35**
	Adequate Vis Ref	STD	
1 & 2 Eng	1/4	1	NA
3 & 4 Eng		1/2	

OBSTACLE DP: Rwy 17, Climbing right turn to 2000'
via heading 200° and TTT R-180 to Nahmu D20.0,
before proceeding on course or AS CLEARED BY ATC.

Some States publish narrative descriptions of their arrivals, and depict them on their enroute charts. They are unnamed, not identified as arrival routes, and are not included in the Jeppesen NavData database. Some States publish "DME or GPS Arrivals", and because they are otherwise unnamed, they are not included in the database.

PROCEDURE TITLES

Procedure identifiers for routes such as STARs, DPs and SIDs are in airborne databases but are limited to not more than six alpha/numeric characters. The database generally uses the charted computer code (shown enclosed within parentheses on the chart) for the procedure title, as

CHART: Cyote Four Departure(CYOTE.CYOTE4) becomes
DATABASE   CYOTE4.

When no computer code is assigned, the name is truncated to not more than six characters. The database procedure identifier is created according to the ARINC 424 specifications.

Database procedure identifiers are charted in most cases. They are the same as the assigned computer code (charted within parentheses) or are being added [*enclosed within square brackets*]. Do not confuse the bracketed database identifier with the official procedure name (which will be used by ATC) or the official computer code (which is used in flight plan filing).

400-FOOT CLIMBS

Virtually all departures in the database include a climb to 400 feet above the airport prior to turning because of requirements in State regulations and recommendations. The 400-foot climb is not depicted on most charts. When States specify a height other than 400 feet, it will be in the Jeppesen NavData database.

Nav2001
AERONAUTICAL INFORMATION NAVDATA DATABASE AND CHARTS

6. ARRIVALS AND DEPARTURES (Cont)

TAKE-OFF MINIMUMS AND CLIMB GRADIENTS
The take-off minimums and climb gradients that are depicted on the charts are not included in the database.

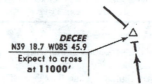

This SID requires a ceiling and visibility of 1200-3 and a climb gradient of 410'/NM to 5000'.

Gnd speed-Kts	75	100	150	200	250	300
410' per NM		513	683	1025	1367	1708	2050

"EXPECT" and "CONDITIONAL" INSTRUCTIONS
Altitudes depicted on charts as "Expect" instructions, as "Expect to cross at 11,000'" are not included in the Jeppesen NavData database. When "Conditional" statements such as "Straight ahead to ABC 8 DME or 600', whichever is later", are included on the charts, only one condition can be included in the database.

DECEE
N39 18.7 W085 45.9
Expect to cross
at 11000'

ALTITUDES
Databases include charted crossing altitudes at waypoints/fixes. Charted Minimum Enroute Altitudes (MEAs) and Minimum Obstacle Clearance Altitudes (MOCAs) are not included. The 5,000-foot altitude at RIANO is included in the database. The MEAs between SURVE and the two VORs are not included.

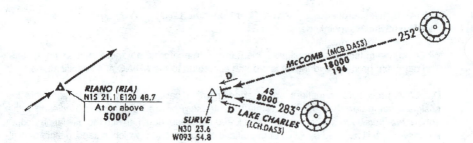

RIANO (RIA)
N15 21.1 E120 48.7
At or above
5000'

SURVE
N30 23.6
W093 54.8

McCOMB (MCB.DAS3)
18000
196
252°

45
8000
283°

D LAKE CHARLES
(LCH.DAS3)

STAR OVERLAPPING SEGMENTS
STARs normally terminate at a fix where the approach begins or at a fix where radar vectoring will begin. When STAR termination points extend beyond the beginning of the approach, some avionics equipment may display a route discontinuity at the end of the STAR and the first approach fix.

Nav2001
AERONAUTICAL INFORMATION NAVDATA DATABASE AND CHARTS

7. APPROACH PROCEDURE (TITLES and OMITTED PROCEDURES)

ICAO PANS OPS approach procedure titles are officially labeled with the navaid(s) used for the approach and are different than approach procedure titles labeled according to the TERPs criteria, which are labeled only with navaids required for the final approach segment. Because of the limited number of characters that are available for the procedure title, the name displayed on the avionics equipment may not be the same as the official name shown on the approach chart.

The Jeppesen NavData database, in accordance with ARINC 424 specifications, codes the approach procedure according to procedure type and runway number. "Similar" type approaches to the same runway may be combined under one procedure title, as ILS Rwy 16 and NDB VOR ILS Rwy 16 may read as ILS Rwy 16. The actual avionics readout for the procedure title varies from manufacturer to manufacturer.

Some avionics systems cannot display VOR and VOR DME (or NDB and NDB DME) approaches to the same runway, and the approach displayed will usually be the one associated with DME.

Currently:

Generally, most Cat I, II, and III ILS approaches to the same runway are the same basic procedure, and the Cat I procedure is in the database. However, in isolated cases, the Cat I and Cat II/III missed approach procedures are different, and only the Cat I missed approach will be in the database.

Additionally, there may be ILS and Converging ILS approaches to the same runway. While the converging ILS approaches are not currently in the database, they may be at some later date.

Some States are using the phonetic alphabet to indicate more than one "same type, same runway" approach, such as ILS Z Rwy 23 and ILS Y Rwy 23. The phonetic alphabet starts are the end of the alphabet to ensure there is no possibility of conflict with circling only approaches, such as VOR A.

In isolated cases, procedures are intentionally omitted from the database. This occurs primarily when navaid/waypoint coordinates provided by the authorities in an undeveloped area are inaccurate, and no resolution can be obtained. Additionally, the ARINC 424 specifications governing navigation databases may occasionally prohibit the inclusion of an approach procedure.

Nav2001
AERONAUTICAL INFORMATION NAVDATA DATABASE AND CHARTS

8. APPROACH PROCEDURES (PLAN VIEW)

INITIAL APPROACH FIX (IAF), INTERMEDIATE FIX (IF), FINAL APPROACH FIX (FAF) DESIGNATIONS

These designations for the type of fix for operational use are included on approach charts within parentheses when specified by the State, but are not displayed on most avionics systems.

ARINC 424 and TSO C-129 specifications require the inclusion of GPS approach transitions originating from IAFs. Authorities do not always standardize the assignment of IAFs, resulting in some cases of approach transitions being included in the database that do not originate from officially designed IAFs

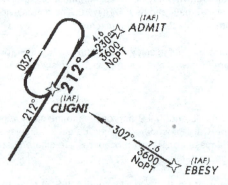

BASE TURN (TEARDROP) APPROACHES

Depending upon the divergence between outbound and inbound tracks on the base turn (teardrop turn), the turn rate of the aircraft, the intercept angle in the database, and the wind may cause an aircraft to undershoot the inbound track when rolling out of the turn, thus affecting the intercept angle to the final approach. This may result in intercepting the final approach course either before or after the Final Approach Fix (FAF).

ROUTES BY AIRCRAFT CATEGORIES

Some procedures are designed with a set of flight tracks for Category A & B aircraft, and with a different set of flight tracks for Category C & D. In such cases, the database generally includes only the flight tracks for Category C & D.

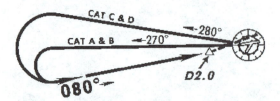

Nav2001
AERONAUTICAL INFORMATION NAVDATA DATABASE AND CHARTS
8. APPROACH PROCEDURES (PLAN VIEW) (Cont)

DME and ALONG TRACK DISTANCES

Database identifiers are assigned to many unnamed DME fixes. The Jeppesen identifier is charted on GPS/GNSS type approaches and charted on any type approach when specified as a computer navigation fix (CNF). Unnamed Along Track Distances (ATDs) are charted as accumulative distances to the MAP.

APPROACH TRANSITION TO LOCALIZER

For DME arc approach transitions with lead-in radials, the fix at the transition "termination point" beyond the lead in radial is dropped by many avionics systems.

West bound on the 22 DME arc, the leg after the 171° lead-in radial may not be displayed in all avionics equipment.

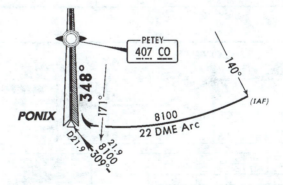

Nav2001
AERONAUTICAL INFORMATION NAVDATA DATABASE AND CHARTS

9. APPROACH PROCEDURES (PROFILE)

VERTICAL DESCENT ANGLES

Vertical descent angles for most *straight-in non-precision landings are included in the data-base and published on charts with the following exceptions:

1) When precision and non-precision approaches are combined on the same chart, or
2) Some procedures based on PANS OPS criteria with descent gradients published in percentage or in feet per NM/meters per kilometer. However, these values are being converted into angles and are being charted.

*Descent angles for circle-to-land only approaches are currently not in the database and are not charted.

In the United States, many non-precision approaches have descent angles provided by the FAA and are depicted on the approach charts. For many of the U.S. procedures, and in other countries, the descent angles are calculated based on the altitudes and distances provided by the State authorities. These descent angles are being added to Jeppesen's charts.

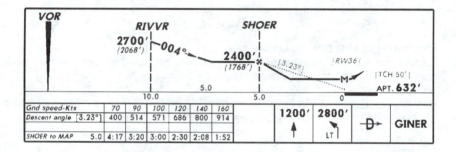

The descent angle accuracy may be affected by temperature. When the outside air temperature is lower than standard, the actual descent angle will be lower. Check your avionics equipment manuals since some compensate for nonstandard temperatures.

Nav2001
AERONAUTICAL INFORMATION NAVDATA DATABASE AND CHARTS

9. APPROACH PROCEDURES (PROFILE) (Cont)

DATABASE IDENTIFIERS
For approach charts where the descent angle is published, all database identifiers from the Final Approach Fix (FAF) to the missed approach termination point are charted in both the plan and profile views. When an FAF is not specified, the NavData database Sensor Final Approach Fix (FAF) is included in the database and is charted.

FINAL APPROACH CAPTURE FIX (FACF)
Databases include (when no suitable fix is specified in source) a FACF for localizer based approaches and those based on VOR DME, VORTAC, or NDB and DME. In most cases, it is the fix identified as the intermediate fix. The FACF is charted only when specified by the State.

GPS/GNSS SENSOR FAF
The Jeppesen NavData database includes a sensor final approach fix when the approach was not originally designed with an FAF, and they are charted on "GPS/GNSS type" approaches.

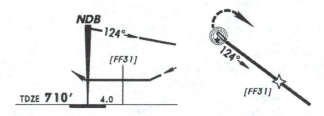

FINAL APPROACH FIX (FAF), ILS and LOCALIZER APPROACHES
There may be several types of fixes charted at the same FAF location - locator, waypoint, intersection, DME fix, OM, or perhaps an NDB instead of a locator. Since many airborne navigation systems with databases don't store locators and NDBs as navaids, a four- or five-character identifier will be used for the FAF on ILS and localizer approaches. The four- or five-character identifier assigned to the FAF location is contained in the waypoint file of the Jeppesen NavData database.

If there is a named intersection or waypoint on the centerline of the localizer at the FAF, the name of the fix will be used for the FAF location.

The FAF must be on the localizer centerline or the avionics system will fly a course that is not straight. Frequently, OMs and LOMs are not positioned exactly on the localizer centerline, and a database fix is created to put the aircraft on a straight course.

When the LOM is on the centerline and there also is a named intersection or waypoint on the centerline, the name of the intersection or waypoint will be used for the FAF. For CHUPP LOM/Intersection, the database identifier is "CHUPP" because there is an intersection or waypoint on the centerline of the localizer at the FAF.

Nav2001
AERONAUTICAL INFORMATION NAVDATA DATABASE AND CHARTS

9. APPROACH PROCEDURES (PROFILE) (Cont)
FINAL APPROACH FIX (FAF), ILS and LOCALIZER APPROACHES (Cont)

When the ILS or localizer procedure is being flown from the database, the four- or five-character
name or identifier such as
CHUPP, FF04, or FF04R, etc. will
be displayed as the FAF.

If the LOM is not on the localizer
centerline, an identifier such as
FF04L may be the identifier for
the computed "on centerline" final
approach fix for runway 04L. If
there is only an outer marker at
the FAF, the FAF identifier may
be OM04L.

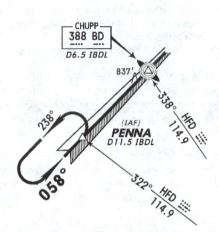

When there is no intersection or waypoint at the FAF such as at the MONRY LOM, the database identifier will be
    "OM09" if the LOM is on the centerline, and
    "FF09" if the LOM is not on the centerline.

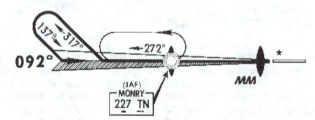

In some systems, to access the locator on most ILS and localizer approaches, the Morse
code identifier can be used.

In the United States, virtually all locators have a five-letter unique name/identifier so the location can usually be accessed in some systems by the navaid Morse code identifier or the five-
letter name. In some systems, the locator is accessed by the name or by adding the letters
"NB" to the Morse code identifier.

Nav2001
AERONAUTICAL INFORMATION NAVDATA DATABASE AND CHARTS

9. APPROACH PROCEDURES (PROFILE) (Cont)

NAMED and UN-NAMED STEPDOWN FIXES, FINAL APPROACH FIX (FAF) to MISSED APPROACH POINT (MAP)

Named and un-named stepdown fixes between the FAF and MAP are currently not included in the databases, but will be added in the future. They are often DME fixes, and in those cases, can be identified by DME. The distance to go to the MAP may be labeled on some GPS/GNSS type charts and VOR DME RNAV charts. Proper identification of these displayed fixes is necessary to clear all stepdown fix crossing altitudes.

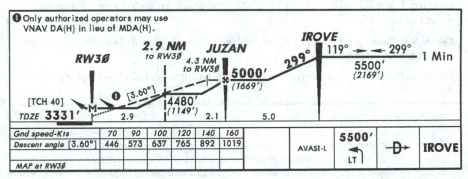

Gnd speed-Kts	70	90	100	120	140	160					
Descent angle [3.60°]	446	573	637	765	892	1019			5500'	-D→	IROVE
MAP at RW30								AVASI-L	LT		

ILS AND RUNWAY ALIGNMENT

Differences in government specified values for localizer and airport variation may cause apparent non-alignment of the localizer and the runway. These differences are gradually being resolved, and whenever possible the airport variation is used for the localizer variation.

10. APPROACH PROCEDURES (MISSED APPROACH)

MISSED APPROACH POINT (MAP)

For non-precision approaches, when the MAP is other than a navaid, there will be a database MAP waypoint with a unique identifier. If the MAP is a waypoint and is at or within 0.14 NM of the threshold the MAP identifier will be the runway number, as "RW04" for Rwy 4 threshold. If the MAP is not at the runway, there will either be an official name for the MAP, or an identifier is provided. GPS/GNSS type approaches, and charts with descent angles, include the database identifier of the MAP.

Nav2001
AERONAUTICAL INFORMATION NAVDATA DATABASE AND CHARTS

10. APPROACH PROCEDURES (MISSED APPROACH) (Cont)

400-FOOT CLIMBS
The database includes a climb to 400 feet above the airport prior to turning on a missed approach. This climb is not part of the official procedure, but does comply with State regulations and policies. This specific climb to 400 feet is not included on charts. The missed approach text supplied by the State authority is charted.

MISSED APPROACH: Turn RIGHT track 080° to intercept CS VOR R-040 (040° bearing from CS NDB). Climb to 5000' and track to D15 CS or GPS or as directed by ATC.

LIMITATION: Max 185 Kt IAS until established on CS VOR R-040 (040° bearing from CS NDB).

CAUTION: Do NOT delay turn onto 080° due to high terrain West of Missed Approach Area.

MISSED APPROACH PROCEDURE
The routes/paths that comprise a missed approach are not always displayed in some avionics systems that use databases. Additionally, some avionics systems that include missed approach procedures don't always implement a full set of path terminators so many legs will not be included in the airborne database. *Refer to the charted missed approach procedure when executing a missed approach.*

MISSED APPROACH: Climb to 1500' then climbing LEFT turn to 2400' via heading 280° and outbound TUL VOR R-238 to KEVIL INT and hold.

11. ROUTES ON CHARTS BUT NOT IN DATABASES

The routes in approach procedures, SIDs (DPs), and STARs are coded into the database using computer codes called path terminators which are defined in the ARINC 424 Navigation Database Specification. A path terminator 1) Defines the path through the air, and 2) Defines the way the leg (or route) is terminated. Not all avionics systems have implemented the full set of path terminators specified in the ARINC 424 document.

Because of the incomplete set of path terminators in some avionics systems, pilots need to ensure their avionics systems will take them on the routes depicted on the charts. If the avionics systems don't have all the routes, or don't have the means to display them, it is the pilot's responsibility to fly the routes depicted on the charts.

FINAL COCKPIT AUTHORITY, CHARTS OR DATABASE

There are differences between information displayed on your airborne avionics navigation system and the information shown on Jeppesen charts. *The charts, supplemented by NOTAMs, are the final authority.*

Nav2001
AERONAUTICAL INFORMATION NAVDATA DATABASE AND CHARTS

GLOSSARY/ABBREVIATIONS

AIRAC - Aeronautical Information Regulation and Control. Designates the revision cycle specified by ICAO, normally 28 days.

ARINC - Aeronautical Radio, Inc

ATD - Along Track Distance, as "3 NM to RW24".

ATS Route - Officially designated route. No designator assigned.

CNF - Computer Navigation Fix

DATABASE IDENTIFIER - Avionics system use only, not for flight plans or ATC communications. Identifies a waypoint or fix.

DP - Departure Procedure

FAA - Federal Aviation Administration

FACF - Final Approach Capture Fix. Database includes (usually as an intermediate fix) when no suitable fix is specified in source.

FAF - Final Approach Fix

FLY-BY FIX - Waypoint allows use of turn anticipation to avoid overshoot of the next flight segment.

FLY-OVER FIX - Waypoint precludes any turn until the fix is over flown and is followed by an intercept maneuver of the next flight segment.

FMS - Flight Management System

GNSS - Global Navigation Satellite System

GPS - Global Positioning System

GPS/GNSS SENSOR FAF - Database fix that changes sensitivity of the Course Deviation Indicator (CDI) on final approach.

GPS/GNSS TYPE APPROACHES - Any approach that can be flown with GPS/GNSS as the only source of navigation.

ICAO - International Civil Aviation Organization

IAF - Initial Approach Fix

IF - Intermediate Approach Fix

Nav2001
AERONAUTICAL INFORMATION NAVDATA DATABASE AND CHARTS

GLOSSARY/ABBREVIATIONS (Cont)

LOM - Locator Outer Marker

MAP - Missed Approach Point

MAA - Maximum Authorized Altitude

MCA - Minimum Crossing Altitude

MOCA - Minimum Obstacle Crossing Altitude

MORA - Minimum Off-Route Altitude

MRA - Minimum Reception Altitude

NavData - Jeppesen Navigation Data

OBSTACLE DEPARTURE - An instrument departure procedure established to avoid obstacles.

PANS OPS - Procedures for Air Navigation Services - Aircraft Operations (ICAO)

QFE - Height above airport or runway, local station pressure.

QNH - Altitude above MSL, local station pressure

SENSOR FINAL APPROACH FIX (FF) - Included in database and on charts when no FAF is specified for the approach.

SID - Standard Instrument Departure

STAR - Standard Terminal Arrival Procedure

TERPs - United States Standard for Terminal Instrument Procedures

VNAV - Vertical Navigation

VERTICAL DESCENT ANGLE - May be established by Jeppesen or specified by the State (country). Charted on Jeppesen approach charts along with database identifiers and rates of descent

WGS-84 - World Geodetic System of 1984

END

The Airline Training Pilot
2nd edition

TONY SMALLWOOD

Ashgate

Aldershot • Burlington USA • Singapore • Sydney

Published by
Ashgate Publishing Ltd
Gower House
Croft Road
Aldershot
Hants GU11 3HR
England

Ashgate Publishing Company
131 Main Street
Burlington
Vermont 05401
USA

Ashgate website: http://www.ashgate.com

British Library Cataloguing in Publication Data
Smallwood, Tony
     The airline training pilot. - 2nd ed.
     1.Airplanes - Piloting - Study and teaching 2.Air pilots -
     Training of
     I.Title
     629.1'325216'071

Library of Congress Cataloging-in-Publication Data
Smallwood, Tony.
          The airline training pilot / Tony Smallwood.-- Updated
          p. cm.
          Includes bibliographical references and index.
          ISBN 0-7546-1161-2 -- ISBN 0-7546-1413-1 (pbk.)
          1. Airplanes--Piloting--Study and teaching. 2. Air pilots--Training of. I. Title.

     TL712.S52 2000
     629.132'5216--dc21

                                                                                   00-29319

ISBN 0 7546 1161 2 (HBK)
ISBN 0 7546 1413 1 (PBK)

Printed and bound in Great Britain by MPG Books Ltd, Bodmin, Cornwall

Contents

7 The brain - memory

List of illustrations

Acknowledgements

The author would like to express his appreciation for all the help and encouragement, received from so many sources, during the preparation of material for this second edition.

Thanks to Robert Helmreich, Ashleigh Merritt, John Wilheim and Robert Gibson, of the NASA Research Centre, Department of Psychology, University of Texas; Paul Baxter, Personal Best Systems; Neil Krey, CRM Developers Group; Captain Ted Murphy, of IFALPA and EUROPILOTE. Dr Ronald Pearson, Head, Medical Research and Human Factors; UK CAA Safety Regulation Group and the many others, who have provided material, the inspiration and the impetus to keep this project going.

Acknowledgements to Lufthansa Flight Training, Oxford Aviation Services, Airbus Industrie, British Aerospace, Flight Safety International, Southern Sydney Institute for providing photographic illustrations. Tom Wise and Dio Tabsila, who have been responsible for the cartoons and drawings.

Introduction

This book is intended as a guide to training objectives and techniques, applicable to all contemporary international jet airliners. The contents should be familiar to all training pilots world-wide. It is hoped that it will provide a valuable insight for all airline pilots, aspiring airline pilots, and indeed all who are interested or concerned with the operation of modern jet transport aircraft.

The demands of technological advances of modern aircraft and their operation have prompted airline management and regulatory authorities to look again at their current training methods. The recognition of the importance of human factors and aviation resource management, as a key to flight safety, has focused the attention on how to improve overall training effectiveness.

How do airlines develop and introduce training programmes for their aircrew? All too often, it is still based on somewhat old fashioned, historically based attitudes and techniques, which do not always prepare crews for complex operations on the modern flight deck. However, times are changing and no commercial pilot or airline organization can now ignore the need for an overhaul of flight crew training methods. In the past decade, a great deal of research has drawn attention to the shortcomings in current flight deck operational procedures.

The aim of this book is to attempt to outline some of the difficulties facing the modern airline pilot, hopefully offer some ideas, provoke argument and further discussion, on this unique and fascinating subject.

What place the pilot?

Airline pilots are highly trained, determined, resolute, dedicated professionals. They are also expensive, ponderous, inconsistent, unpredictable, prone to mistakes, suffer stress and fatigue, need eight hours sleep every day and are relatively ineffective and poor monitors of systems. Modern digital electronics on the other hand are relatively cheap, accurate, fast, compact, and predictable and are able to show intelligence. Putting pilots in aeroplanes creates the need for a costly complex man/machine interface, and yet nearly 80% of all aircraft accidents can be attributable to avoidable human error.

What is going to keep the pilot in his job?

Given the phenomenal improvement in computing software and capacity, in the last decade, the only thing that is stopping the introduction of unmanned airliners is customer acceptance and current specifications for all new aeroplanes. If pilots are to retain their role, and avoid being phased out, like their flight engineer and navigator colleagues before them, then they, and company training departments, most certainly have to put their house in order and take steps to reduce this human error.

Human factors

Studies on human performance presented at a seminar, held by the Royal Aeronautical Society in 1998, echoed findings by Boeing, which concluded that many accidents and incidents could have been prevented by pilot adherence to published procedures.

How can pilots avoid making costly errors?

Training, whether initial or recurrent, is one of the more traditional ways to reduce mistakes, by setting out to instil the disciplines that make rule breaking or disregard for *Standard Operating Procedures* (SOP's) less likely. Training has to do more than impart skills and knowledge. It has to create a level of consciousness, which enables the trainee to recognize problems met in the real training environment.

One traditional method of preventing mistakes is punishment. Implying that safety could be improved by enforcing procedures more ruthlessly. The proof is that this has not always worked in the past. Perhaps it is time to examine both the traditional training process and the psychology of deterrence.

Over the past 5 years, a big rise has been recorded in crew errors, resulting, either from insufficient pilot knowledge of aircraft systems and procedures, or pilot proficiency failure. The aircraft procedures and systems knowledge deficiencies, classified as 'H3' errors by the International Civil Aviation Organisation (ICAO), have been seen to rise fourfold in aircrew flying aircraft with highly automated flight decks.

The causes suggested by industry studies, focus on the complexity and *the failure to adapt training appropriately* to the needs of pilots flying modern aircraft. Taking 1998 alone as an example, there were 13 'loss of control' accidents in which 364 people died, compared with 1997's figure of 9 and 257. The International Federation of Airline Pilots' Associations, amongst other agencies, have voiced their concern about the increase in numbers of this type of accident over the past decade.

Many accident reports indicate that professional pilots often neglect the human input, by not considering the *softer* issues of flying, that are those related to crew interaction, decision making, leadership and resource management. One of the most demanding roles of the airline training pilot is in this area of human factor management and assessment.

Increasing technical reliability and safety of aircraft has left human factors crew training, and flight management, in its widest sense, as the main areas in which significant advances in flight safety can be made. Accidents caused by technical faults represent 3% to 4% of all cases, human error, as already seen, close to 80%. This figure is showing an alarming increasing trend, as current accident statistics show, as just ten years ago human error accidents were closer to 70%.

There is concern in the industry that regulations governing pilot conversion training for the latest generation airliner flight-decks are inadequate, and have not kept pace with the problems of new pilot/machine interfaces presented by these aircraft. A number of recent accidents have demonstrated that there has been failure of 1990s operators and regulators to recognize the magnitude of the differences, between 1970 and 1990 aircraft, and the respective ways they should be operated. Different training is required for a pilot making the transition from a traditional cockpit to a modern EFIS one. It is not that these new flight decks are necessarily badly designed or more difficult to use, in fact overall they are easier, but they are different and require different training emphasis.

In general, it is felt that the aerospace community has not faced up to the importance of that difference. Old style conversion is no longer adequate to meet these new needs.

Training pilots on modern airline transport aircraft, on line revenue operations, is a very challenging task. Most pilots selected for the role of training staff enjoy the work, and if the task is properly approached a new and satisfying dimension can be discovered. As well as helping other people and making a tangible contribution to the industry, a chance is offered to give something of themselves, in return for all that has been instilled and invested in them over the years. One of the airline's greatest resources is the accumulation of skills and experience of its training staff. We must continue to use this human resource to the best of our ability. Although the financial investment in training is huge, the real quality of the product still depends on the individual.

We already expect good *stick and rudder* skills, from our airline pilots, as well as an increasing accent on flight management abilities. There is little doubt with ever-increasing automation, on present and future jet airlines, that developments of flight management skills are going to require a greater emphasis from training departments. The problem is in maintaining a good balance. The advancement of avionics is irreversible. The challenge is to ensure that pilots never become bored screen watchers. Pilots need to be kept in the decision making loop.

Error management

The inescapable fact of life is that human error is inevitable and ubiquitous. Error can never be completely eliminated but what the pilot can do is:

- Reduce the likelihood of error;
- Trap errors before they have operational effect;
- Mitigate the consequences of error.

Trainers and evaluators need to expand their role to train and reinforce recovery from error and the management of inevitable errors. A 'new' approach is required, concentrating on assessing and reinforcing error management strategies.

For the foreseeable future, the human operator remains central to safe and reliable aviation activities. Therefore, the importance of addressing human related error remains critical to maintaining and improving safety.

Crew resource management

One of the biggest changes, during its 20-year history, that has affected airline pilot training world-wide has been the acceptance of crew resource management (CRM); called by a number of different acronyms in the past but the industry seems now to have settled for CRM. CRM is now mandated training in many countries for pilots, and in some cases, flight attendants, maintenance personnel and flight dispatchers. CRM has evolved *(and is evolving)* taking on board differences both of organization and national cultures and widening healthy debate.

CRM

The operation of a modern transport aircraft in today's aviation system is a complex undertaking that requires a cohesive team, in the air and on the ground, to accomplish its goals. In common with machinery, human performance has thresholds and upper limits beyond which performance is compromised. Given the complex and demanding working environment error can never be completely eliminated. Recognizing the complexity of the task and the limitations of human performance allows greater safeguards to be inserted into the system. CRM is just such a safeguard.[1]

Airline and pilot attitudes to CRM have ranged between scornful rejection, passive adoption and unswerving acceptance, but the aviation psychology community asserts that much of the potential human factor activity has yet to be developed and exploited fully. However, the growing international dialogue, and application of the human factor, is measurably improving air safety. New initiatives can now reach far beyond early CRM philosophies into company-wide safety system enhancement through error management principles and a better understanding of human frailties and cultural differences.

Getting the edge

There are 90,000 professional pilots in USA (1996). 60% are employed by airlines and 95% are male. Over 60,000 airline pilots will be hired in the next 10 years.[2]

During the 1980's 15,000 pilots in USA were scanned to fill the first 80 slots for major carriers. The odds remain the same today. If you are going to be a successful pilot for a major carrier then you are going to need *an edge* over your fellow applicants. Indeed, if you are to be successful in your career as a pilot, then a sound knowledge and understanding of the principles involved in airline flight crew training will undoubtedly enhance this process.

Conclusions

Aviation has the respect and admiration of the flying public for its extraordinary safety record. It has reached this level of success by identifying and pro-actively addressing safety issues at all levels and across all components of the aviation system. However, with this success comes a tremendous responsibility to not just maintain this level of safety, but also improve it further. Whilst not wishing to denigrate the wealth of experience

---

[1] Merritt, A.C., & Helmreich, R.L. – CRM! I hate it! What is it? Paper presented to the Orient Airlines Association Seminar, Jakarta April 1996.
[2] Source: "Becoming an Airline Pilot" Griffin, J.

and expertise built up over the years, the fact remains, if we are to reverse the trend of pilot error induced accidents, then airline training departments and airline management need perhaps to reappraise their methods. What was adequate and satisfactory ten years ago may not now be the case. The evidence is that we have not yet got it right.

It is hoped that this book will show how to overcome some of these problems, by the pursuit of more effective and efficient flight training. Where we have used the word 'he' or 'him' in the text we have deliberately omitted 'she' or 'her' purely in the interests of word economy.

Note: A comprehensive list of related web sites is included on pages 330-336

1 Where are we now?

Concorde has already been crossing the Atlantic for over thirty years. With extensions to the aircraft's life span, it may last into the beginning of the next century. What then? It seems likely that once Concorde retires then there will be a gap of several decades before the next generation of supersonic transports (SSTs) emerges. These successors are likely to be hypersonic aeroplanes with all the advantages of high altitude and high speed. Second-generation jet transports, still in service, like the B747, B737, and DC-9 have all been around for the same number of years. Derivatives of these successful designs are still being produced, with obvious design improvements. These are regarded as the third generation of subsonic transports, and include types like the B777, B757, B767, B717, A320, A330, and A340. All these aircraft incorporate *glass* cockpits.

The future

The next couple of decades are going to see the fourth generation of subsonic design derivatives of present types. With further advancements in automation, economy and efficiency, it is likely these types will change little, externally, from current designs, but what will change are the traditional pilot concepts and terms of reference. *Flying* the aircraft will no longer be the pilot's primary task. Instead, he will be at the centre of a communications network, monitoring and supervising the overall situation. Automation should improve to reduce the pilot workload so that he can spend more time coping with the increasingly complex congested situation outside and inside the aircraft.

      The next big quantum jump in aircraft design will be the arrival of the mega-bodied 600-1000 seat aircraft. This will make even larger demands on pilot responsibility. It is likely that some of this responsibility and control will need to be shared with ground based operators and controlled by secure data links. These data links to flight management systems (FMS) will offer many operationally attractive advantages, doing away with the almost pre-historic R/T chatter, allowing aircraft closer separation, thereby improving traffic flow. The emphasis of design must be to give computers and pilots that which they are best at doing, leaving the pilot as the ultimate decision-maker.

These designs will almost certainly feature *paperless* cockpits, they will have an integrated documentation system with electronics that remove the need for pilots to juggle with maps, charts, and approach plates. The aircraft will carry all the data in its electronic library, updating by databases from the ground. Cockpit design has undergone drastic changes since the introduction of computer controlled electronic flight information systems (EFIS). Modern day jet airliners bear little resemblance to the previous generation of analogue displays.

The captaincy question

The modern flight deck is a multi-task environment. The flightcrew are constantly faced with multiple, concurrent, competing, often conflicting goals to accomplish, and therefore must engage in multiple activities to accomplish them. As most pilots are aware it is not only difficult to successfully accomplish such goals, it is even more challenging to manage the activities directed towards them.

Employing and understanding task management is seen as a major player in reducing 'pilot error' accidents. Some of these errors are in performing flightdeck functions, and others are in managing flight deck goals, and the functions to achieve those goals.

Airline pilots have gradually come to accept that they will relinquish an increasing amount of control to onboard computers. The next and more controversial step is how far airlines should accept intervention from systems outside their aircraft. Technology already exists that allows real time monitoring and even control of an airliner from the ground. The future airline pilot's task will be an integral part of a management system that is linked via digital data link systems giving ground controllers and airline operators real time checks on flight deck activity. We will need to appreciate that the pilot's task is becoming ever more closely monitored and controlled.

Culture

In aviation training, the three cultures, professional, organizational and national, have both positive and negative impact on the conduct of a safe flight. Safe flight is the positive outcome of timely recognition and effective error management, most definitely universally desired outcomes. The responsibility of training departments is to minimise the negative components of each type of culture, whilst emphazising the positive.

Application of CRM principles, coupled with sound technical training and a positive organizational structure is an important part of error management philosophy.

The 4th dimension

Passenger traffic is forecast to double by the next decade. Air traffic control centres, for example, in busy European airspace, are predicting that their systems will eventually have to be computer linked to the aircraft flight management system (FMS), via their own 4th dimension navigation computer, the 4th dimension being time. This will allow aircraft to fly according to electronically filed flight plans, with accuracy, impossible to obtain using manual techniques. The human factor implications of this are fearsome.

Questions

During preparation of material for inclusion in this book, a questionnaire was distributed to a number of airline crews. Questions were asked on the subject of training in order to determine some attitudes towards training problems and provide some practical input, which would be of assistance to present and future training captains.

The response generated was very positive and a great deal of useful information was returned. This response reflected in some way the amount of training activity, which had taken place during the previous two years, with new aircraft types being introduced. More particularly, because of the numbers of pilots who became actively involved in the project, it showed the great interest in training matters which aircrew have generally. This is brought about by the on going nature of the profession we are either training or being trained almost continually. Nothing is static for long. Difference and change are always just around the corner. We are all involved.

Training

Questions were asked on attitudes and usefulness of ground training aids and supernumerary flying exposure.The answers revealed that the available training aids had not been as well used as could have been expected. This is partly due to a re-education process, which is still going on. Pilots have to appreciate that the *old days* of rather cursory learning on the ground, with the attitude that, *I didn't get much from the course but no need to worry, I'll learn it when I get into the aeroplane,* have gone.

It has been found conclusively that, from both cost and efficiency points of view, pre-flight learning using ground based training aids, including full simulator endorsement time, produces a better trained pilot. The aids will do the job. The problems are still in access, availability and attitude. The approach has to be well organized and sequential, with plenty of follow up time for practice and consolidation. Study and application are still just as necessary on the part of the individual, but because the learning is so much *hands on* and learning by doing, it is much more effective. Most learning from now on will be done in a cockpit environment whether audio visual trainer, or simulator. The overall training time, particularly the transition from one aircraft type to another will be reduced. Some psychological barriers have to be overcome, for example, sitting at a teaching machine console, with less time spent listening to, or talking with, an instructor. However, short periods of such activity can be more beneficial than long periods spent in classrooms.

Ground training

- Acquisition of engineering knowledge of the aircraft
- Application of this knowledge on the cockpit
- How do things work normally
- How do things work non normally
- Acquisition of knowledge of sub-systems
- Integration of the aircraft systems and flying procedures
- Normal, non-normal and emergency flying procedures, aircraft endorsement (simulator).

It is still a fact that the flight simulator does not reproduce *all* the illusions of flight. A great deal of learning can still be accomplished in supernumerary flying.

Supernumerary flying

In answering the questions on this subject, most trainers felt that their trainees were not obtaining maximum benefit from supernumerary flying. Some had done none at all. As well as better integrating the learning achieved during ground training, a great deal of time can be saved by some effective supernumerary flying. Time spent on line training can therefore be more productive. It was felt that the trainee should undertake an *observe and learn* process during some supernumerary flying watching line crews operate the aircraft. During this time, he should be guided by a supernumerary pilot's schedule.

At the completion of ground training and supernumerary flying the student is in a position to gain maximum benefit from his line training programme.

Which areas of training demand the most attention from the training captain?

Subjects marked * are the most likely areas of potential difficulty:

- Flight planning
- Conduct of good briefing*
- Inspections and pre start
- Taxing and ground handling
- Radio procedure
- Support duties*
- Flight management*
- Systems management
- Take off profile and initial climb*
- Climb and en route instrument flying
- Automatic flight control system*
- Performance data computer system
- Flight management system
- Use of navigation aids
- Prop to jet conversion*
- Initial intake training*
- Initial command training*
- Turbulence and windshear techniques
- Descent profile*
- Final 30 miles*
- Instrument approaches*
- Circuit flying: Day/Night, wet runway, crosswind*
- Final approach flare and landing roll*
- Reverse thrust, braking techniques.*

The line training programme: What are the problem areas?

Pilots were given a list of training tasks, and asked to indicate the areas with which they had encountered the most difficulty during their line training. They were further asked to indicate whether they thought their own preparation had been the main factor in this difficulty, or whether it reflected more the quality of the training that they had received. Training captains were also asked to comment on the same list of training tasks and to indicate perceived problem areas.

The answers revealed two prime considerations:

- There is a very consistent pattern of problem areas in training regardless of aircraft type

- There is a need for improved information and guidance for trainers.

The areas of difficulty were as follows:

- Acquisition of high standards of flight management skills, both in first officer and command areas

- Acquisition and maintenance of instrument flying skills

- The descent profile, and the last 30 nm in particular

- Maintenance of skills required in local asymmetric training in both simulator and aircraft.

These factors indicate that there is much more to a training pilot's job than just teaching his trainee how to fly the aircraft. On intake and initial command training in particular, he is regarded as a source of information on a variety of subjects. In all forms of training, he is regarded as the model and tutor. It is clear that line pilots under training demand a great deal of skill and expertise of the trainer.

Two further problem areas emerged with particular reference to intake pilots:

- Orientation and general airline aircraft operation

- The concept of first officer *support* duties.

These areas require special attention and understanding on the part of the trainer. He should not assume too much in the early stages and be prepared to instruct and teach in many areas of airline operation, with particular stress on standard operating procedures as defined elsewhere in this book.

Some short-term learning problems were indicated in the area of advanced cockpit technology.

It is interesting to note that basic aircraft handling appears to be a problem in the latter stages of descent and in instrument flying procedures in all aircraft types. Trainers need to pay particular attention to these areas. Guidance should also be given in preparing a trainee for the requirements and techniques involved in local asymmetric and simulator exercises.

A training challenge

In 1994 the US National Transportation Board (NTSB) conducted a study and found that 80% of 37 accidents examined occurred while the First Officer was functioning in the pilot-not-flying (PNF) role. The study also found that over half of the F/O's involved had less than one year's experience.

A further study, completed by the University of Central Florida in 1997, suggests that F/O PNF difficulties are not typically related to technical or skill problems, rather they stem from the limited ability to manage workload effectively.

As two-pilot crew aircraft are now the norm in most airline fleets, new F/O's are entering airline fleets who have less experience than in the past. These inexperienced pilots are considered satisfactory in their technical ability with regard to pilot flying (PF) duties, but often have not had the training or experience performing as PNF.

These studies suggest there are performance requirements for F/O's that many training programmes do not currently address. The major problem identified was the pilot's failure to monitor and challenge the captain. This is attributed to an inability to plan their own workload. Although identifying that a problem existed, they did not initiate action, either because they believed the problem did not require corrective action at that time, or did not act for fear of potential negative consequences.

Specifically new F/O's require training in monitoring and challenging, within the context of CRM and line orientated flight training (LOFT) programmes. Initial training programmes should ensure that the newly qualified F/O is given ample opportunity to perform the duties of PNF. Trainers can enhance the F/O's early development through case studies, discussion and exercises, to recognise the skills and issues involved, including the conduct of post flight de-briefing, which focus on in flight performance.

Problems indicated on specific aircraft types

British Aerospace ATP

Considered a *very different* aircraft on first acquaintance, but later very satisfying to fly and an excellent training aircraft. First Officers making the transition from jets onto the ATP for command need special attention from the trainer. Basic airmanship needs to be re-emphasised.

DH DASH- 7

Considered a very versatile and satisfying aircraft to fly. Special attention from the trainer in the areas of STOL performance - high angle approaches and possibly difficult and unusual airfields.

Fokker FK70/100

A straightforward and very pleasant aeroplane to fly. Good handling characteristics with good field performance and versatile descent/speed control, make it a relatively easy aircraft to fly, presenting few problems for pilots converting to type. High technology automatic flight capabilities, as in most modern aircraft, reduce the time available for a pilot to manually practise. This problem needs to be addressed by trainers. Complex, state of the art, EFIS/FMS and electronics will need careful attention, so that trainees are fully conversant with all modes of operation.

Boeing 737-300/400/500

Handling is typical of a light/medium weight swept wing jet transport. Because of the low-slung engines, the thrust/pitch change inter-relationship is quite marked and trainees will need to anticipate this. Operation of the aircraft inside 30 miles is probably the most difficult task for non-jet pilots upgrading to type. The trainee should be made aware that the aircraft is flexible enough to accommodate most ATC requirements relating to speed control. However to achieve significant speed reduction, the aircraft needs to be flown level, with speed brake, where appropriate. Lack of understanding of the aircraft's characteristics can result in fast approaches that make a stable approach gradient difficult to achieve. Overall considered an easy aircraft to fly and much easier to land than the heavier types. Problem areas lay in automation, teaching pilots how to use the advanced technology built into the aircraft and yet retain a high standard of manipulative skill.

The trainer on the B737 is very busy because of the short sectors and the amount of new equipment that the trainee has to absorb in the time available.

Boeing 717 (MD 80/90)

Considered a delightful aircraft to learn to fly, and very satisfying to operate after some experience on the type. Problem areas for trainers include:

- Flight management for new captains and first officers

- Flight planning, long range operation, and flight performance

- Appreciation of large trim range with the large variation in aircraft operating weights, and C of G range

- The approach, landing, flare and touchdown.

Note: This last item was considered a problem requiring special attention from trainers for all trainees, regardless of experience on other types.

Boeing 757/767

No particular problem areas emerged, especially from pilots who had experience on other Boeing types, other than the adjustment to advanced cockpit technology. Most pilots considered it an easy aircraft to fly after some experience.

Boeing 747-400

A remarkably efficient renewal of a 30-year-old design. The original flight deck felt crowded and busy, having 971 lights, gauges and switches compared with 365 in this version (22 less than the 757/767). The two-man flight deck gives an easy transition from the original *Jumbo*. The usual question is *how did I manage before!* Its raw performance and responsive handling give little hint of its colossal size, having a remarkable 60°/sec roll rate.

Trainers need to stress the importance of thinking ahead during the initial stages of the approach, to slow down and select flaps in plenty of time, whilst keeping a watchful eye on the descent profile. Speed stability is rather low on the approach, needing only very small pitch or power changes for big results.

Perhaps the biggest problem facing airline training departments, is maintaining current recency and practice for its crews. Long sectors (13hrs+) and the use of relief crews en-route, means that pilots do far less approaches and landings than they used to.

Airbus A320/319/321

With its fly by wire, side-stick control and all-composite fin and tailplane, the A320 and its derivatives, represented something of a technical revolution. Commonalties between the types, reduce training costs and conversion times improving productivity. Conversion training between types needs special consideration. The A321, for example, has a 6.93m fuselage extension requiring a 2° less rotation than the A320, to avoid tail strike. Pilots routinely flying these variants need to be aware of the differing techniques.

Perhaps the biggest problem confronting trainers is to ensure that their trainees are fully conversant with the aircraft's advanced automation and control laws. Much has been written elsewhere in this book, on this subject of man/machine interface.

Airbus A330/340

The newest long haul, wide-bodied pair. Common cockpit layouts across the whole range of Airbus aircraft have enabled cross crew qualification (CCQ). Pilots can be switched between types without retraining. Airbus Industrie have pursued the aim of having them feel identical to the handling pilot, even though, for example, the A340 has four engines and is 3½ times the size of the A320.

Although handling and flight deck layouts may be very similar, pilots need to be aware of different performance requirements, different wingspans, when manoeuvring on the ramp, and different cockpit heights. When judging flare, for example, the runway looks much smaller from the A340. Nearly all features in the A340 are indistinguishable from the A320 despite the greater weight. The A340 does however appear *softer* to fly than the A320. A ten-day conversion from the A320 to the A340 is considered normal.

Boeing B777

The world's largest twinjet airliner and Boeing's first aircraft using full fly by wire (FBW) electronic controls. Innovation in cockpit design, advanced options in data presentation, emphasis on ergonomics and human factors make it particularly user friendly. Improved pilot situational awareness, through better graphic displays, reduced head down time in flight management system (FMS) operation, and an aircraft information management system (AIMS) are a big step in technological advances.

This aircraft represents a big change in Boeing's design and production philosophy. FBW allows the use of lighter wing and tail structures, replacing complex mechanical cables, brackets and linkages. The company maintain that FBW will be more reliable, easier to maintain, and offer better handling characteristics. The Boeing system *has functions to assist the pilot in avoiding, exceeding operational boundaries.* Conventional operation is with *normal* stick and control wheel, the FBW being *invisible* to the pilot. Mechanical reversion is possible in case of overall FBW system failure by an independent mechanical link controlling the stabilizer trim system and flight spoilers. This aircraft is now in worldwide airline service and will no doubt present further challenges.

Airline Pilots' Association

The US Airline Pilots Association (USALPA), all weather flying committee recently reported that *the aviation system is not optimized for human operations, it requires constant human intervention. The efficiency and usefulness of current automated systems drop out rapidly as conditions become less predictable.* In other words, it is often when the FMS is most needed that it is the least helpful.

Several years ago the National Aeronautics and Space Administration (NASA) published an extensive study *Human Factors in Advanced Technology Aircraft.* The results indicated that advanced automation could and does decrease workload during periods when it was already low, but could increase it in a high workload situation.

A study of the implications of aircraft automation

The level of flight deck automation has increased dramatically over the last 20 years. With current available automation, pilots can delegate just about all flight control and navigational tasks to computer controlled systems.

Although the overall accident rate has decreased since the widespread use of automation, this overall decrease masks the existence of a trend in accidents and incidents that may be partly or wholly due to pilot's interaction with automated systems.

It has been suggested that flight management computers (FMC) may offer too many possibilities and be too complex, with the result that many pilots rely on only 20% of the software features.

Experiments have shown that automation use can lead to excessive trust and deceased vigilance. For example, there were 248 reported level busts in the UK flight information region (FIR) in 1997. How many more went unreported! How many minor altitude deviations occurred that could have resulted in a level bust!

Research[1] indicates serious limitations to human performance with regard to automation and presents a challenge not only to training departments but also the trainee. With the wide range of options for control of flight, pilots are afforded considerable discretion in their use of automation. Ultimately it is the crew's decision, and standard operating procedures (SOP's) as to how and when to employ it.

Technology

Current standards for type certification and operation have not kept pace with changes in technology and increased knowledge about human performance. Accidents are occurring because many pilots do not understand their aircraft. Lack of comprehension, confusion, poor knowledge of how to interpret system information presented on displays, and ignorance of the interfaces between autopilot, data sensors and control services, have been the direct cause of several fatal hull losses.

Accident report

> The pilot deciding that the aircraft was too high on ILS approach (about 1,300 ft with 2 km to threshold) began a go-around. During the go-around the aircraft climbed rapidly to 2,750 ft, adopting a maximum pitch up attitude of 42.7°. During the descent the aircraft banked dramatically left and right and attained a pitch down angle of –44.6°. The pilot regained control too late to recover and the aircraft hit the ground tail first to the left of the runway. 196 fatalities.[2]

---

[1] National Culture and Flight Deck Automation: Results of a multi-nation survey 1997. University of Texas at Austin. Sherman P., Helmreich R.L., Merritt A.C.

The recent B757 Cali accident was a case of programming, and failing to check the flight management system (FMS) which, after take-off, promptly turned the aircraft the wrong way, flying into the side of a mountain.[3]

Further examples of pilot failure to interact with automation, caused the loss of the Burgenair B757 in February 1996, and the CAL accident at Nagoya. Almost a mirror image of the accident four years later.[1] Pilot error, pilot confusion, was different in each case, but all these accidents involved pilot ignorance of various aspects of the automated flight control system. The pilots involved showed lack of comprehension of what the aircraft was doing and why.

Currently the concept of teaching generic automatic flight system principles, highlighting what the system does well, and when its use should be avoided has historically been low on the list of instructional priorities. Introducing automation-specific human factor issues, such as mode confusion, situational awareness and highlighting human performance shortcomings need to be re-balanced within training programmes.

The International Air Transport Association (IATA) 5 years ago undertook a study[o] of the problems that result from operating modern *glass cockpit* aircraft. The idea was to attempt to gain some advanced warning of the safety problems that were developing, by studying a large number of incident reports, where questions could be raised in the use of automation, software design, and the integration of human capabilities into aircraft operation.

Conclusions

- There is a problem with complacency and total acceptance of output of the Flight Management System (FMS)

- FMS equipped aircraft crews do not adequately monitor basic flight instruments

- FMS can lead to increased workloads at critical stages of flight

---

[2]China Airlines (CAL) A300-600 Feb '98, Chiang Kai Shek
[3]American Airlines (AA) B757 Dec '95 Cali, Columbia
[1]Burgenair B757 Feb '96 Puerto Plata, Dominican Republic. China Airlines (CAL) A300 April '94 Nagoya, Japan
[o]IATA Aircraft Automation Report. Implications of Aircraft Automation July 1994

- The increased number of modes available for autopilot and autothrottle use present situations not always understood by the pilot

- Current training methods do not stress the importance of basic flying skills, nor provide the opportunity to practice flying with minimum use of automatic systems

- Aircraft technical manuals are incomplete or cover subjects in too shallow a detail

- Design targets are not met. The desire to reduce pilot workload has merely shifted the workload

- Cathode Ray Tube (CRT) displays have increased information available. Different monitoring techniques are required

- The interaction of Automatic Flight System (AFS) components does not always clearly indicate when one or other has been disconnected or for what reason.

The next chapter investigates the role of the trainer.

A shortage of highly qualified candidates is likely within the next few years.

2 The training pilot

Perhaps the ingredients most required by a training pilot are common sense, enthusiasm, flexibility, patience, a keen sense to observe and a strong sense of humour. As in most pursuits, an organized and well planned approach will make the training task much easier.

The superstar teacher/trainer

The ideal trainer will exhibit some of the following attributes:

- Vitality
- Energy
- Vigour
- Wit
- Optimism
- Seriousness
- Wisdom
- Rapport
- Experience.

He or she will be expected to stay cool, manage stress, practice principles, show empathy, be competent and professional, alert yet relaxed, create and meet high expectations and possess good presentation skills.

Preparation for the training pilot role

To be selected as a training pilot you not only need perceived and demonstrated skills in aircraft manipulation and management, but an ability to impart knowledge, along with a mature personality, and practical abilities in managing people. Somewhere along the way, you will be called upon to demonstrate these qualities. Hopefully not all at once!

The situation where the overload, and preoccupation with the trainee syndrome, has the potential to create a breakdown or weakness in the training pilot's own flight management is, with foreknowledge, to be avoided. This potential is probably at its greatest when the trainer himself is relatively new to the task and is given charge of a new trainee. You constantly need to maintain the awareness that your primary task is to conduct a safe and on schedule operation, that is comfortable for the passengers and economical for the company.

Training categories

A trainee pilot on type can be one from any number of categories:

- Initial intake

- Initial command

- Conversion   Prop To Prop
                  Prop To Jet   - may also include initial intake
                                     and initial command
                  Jet To Prop
                  Jet To Jet

- Requalification e.g. following an extended leave period

- Retraining e.g. second training period following failure to qualify at first attempt - may include initial intake, initial command or conversion

- Extra training e.g. failing to clear to line with only minor difficulties or unsatisfactory licence renewal

- Revalidation e.g. pilot out of recent experience.

Each of these categories requires special consideration and sympathetic understanding by the training staff. Consider the case of the initial intake or command pilot! Either would approach the task with high morale and great enthusiasm, whereas in a retraining situation the pilot will most likely be suffering from low morale and lack of confidence. In this latter case you will be confronted with your greatest challenge, to help rebuild the trainee's confidence. Conversely, an overconfident approach can usually be readily corrected, the danger being however, that in some cases the overconfidence may mask an underconfident psyche and overcorrection could create greater problems.

We want our trainees to be:

- Competent
- Safe
- Efficient
- Achieving personal satisfaction from the task.

It is interesting to note that if the fourth factor above is not present, then the first three factors are all affected. The training pilot becomes a very important influence as a model of both skills and attitudes, and is the main source of encouragement for the trainee. You not only want your trainee to be successful as an individual, but it is necessary for him to be able to function as part of a team, upholding the philosophy of standard operating procedures, which have been developed over the years, often through hard won experience.

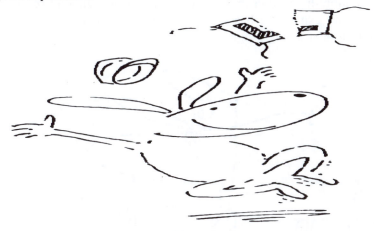

...achieving personal satisfaction from the task!

Standard Operating Procedures - SOPs

To enable crew members to operate the aircraft competently, safely and efficiently, recognizing the authority vested in the captain, as well as individual responsibility, to exercise established company procedures, with co-operation, correct habit patterns and predictable actions, no matter what circumstances or personalities are involved.

Within the context of this definition, crew members are expected to display initiative, enthusiasm and self-reliance. Good standard operating procedures are not easily achieved and must be constantly strived for, monitored and reviewed, in the light of the many variables present in such a dynamic environment. Training pilots play a very important role in implementing these high ideals that have such a significant effect on the successful operation of the airline.

Trainer v trainee relationship

Successful training depends largely upon the establishment of a satisfactory relationship between trainer and trainee. Arrange to meet your trainee on an informal basis. This will usually invoke a positive response that will work to the advantage of both. Remember that first impressions will be made on both sides. Be yourself, do not attempt to act out an expected role. Avoid the error of either a *talk down to his level* or a *dictatorial* approach. Meeting and talking things over involves both parties in giving up some of their own time, but experience has shown that it is well worth doing both before and during the training programme.

...the dictatorial approach!

The trainee

At this first meeting, your student will often be found, depending on his/her own level of experience, to have emotions that contain elements of being keen to start, tired after a preceding stage of school/simulator phase, overwhelmed by the task ahead of him, confused about which book to read first and where to start, doubtful as to *whether he can make it* particularly if he has had trouble with earlier aspects of the programme, and so on. The over confident trainee may have little appreciation of what is ahead of him, particularly in the allocation of his time in the training process, and the subsequent priorities which will have to be established in pursuance of some of his other activities. He may well have the attitude: *Well here I am, now it's up to you to get me through.*

Your attitudes

This early meeting can be a most useful experience. You will be commencing the process of adapting to your trainee and yet not changing your own basic personality or your resolve. Your trainee needs to see some of your attitudes. Let your enthusiasm show. Be positive. Compliment him for the success achieved so far, with the theory, simulator and type rating programmes. Talk about the aeroplane, and the enjoyment and satisfaction in flying it. After all, to most pilots, flying is still controlled, disciplined *fun* if approached with a positive mental attitude. There is even a school of thought that says that if flying is no longer *fun*, it is time to stop flying.

…have your trainee appreciate you at all costs!

You must not allow your friendly approach to be interpreted as being *over eager, trying too hard* or being determined to *have your trainee appreciate you at all costs*. Rather, you must stress that you are interested in your trainee. You are there to help. You take pleasure in doing the job properly, which involves adherence to high standards of manipulation, procedural knowledge and flight management. The often-stated guidelines of being firm, fair and friendly are most applicable.

Other attitudes need to be discussed. The standard of dress and presentation as an airline pilot is important. You take pride in your appearance and you expect him to do likewise during your period of training association.

Remind him that the training workload is heavy in the time available. On overnight stops, this will necessitate a common sense approach towards social drinking. Good meals and plenty of sleep are necessary. Of course, most pilots do not need guidance in these areas, but occasionally an individual will make life unnecessarily difficult for himself, and a timely word from you may be appropriate. Emphasize that he will need to do some work at home in the form of preparation and learning, but that time still needs to be allocated to all the other demands of life. Help him to determine a sense of balance and a programme in this area. Relaxation and exercise are important to both parties during the training programme.

Supply your trainee with a copy of the complete training form, listing all requirements, before commencing to talk about your intended plan of approach, suggested reading, and starting point. Talk about your trainee's past flying experience. Determine the degree of preparation undertaken for the new programme. Has he done the required amount of supernumerary flying? Has he made good use of all the training aids available? If there is doubt about the completeness of his preparation, advise him to attend to the deficiency immediately before training commences.

Some guidance in this area is appropriate. Having just come through the processes of school, theory, simulator and type rating, he has flown the aircraft and achieved a certain standard. It is reasonable to expect that he will have a fair knowledge of the aircraft limitations and emergency drills. Most importantly, he should know how to operate the oxygen and the associated communication switching. This aspect will require a thorough check by you at the start of the training. Do not expect too much too soon.

A working knowledge of scan patterns, internal and external checks lists is desirable. It has been found that mutually designed *prompt* cards are very effective learning tools for these areas when used early in the programme.

The trainer too, should carry a notebook during the training period, in which to record relevant comments and notes on progress. This technique will prove to be most useful to you, both as a reminder and as an aid to further training. Inform the trainee that you will occasionally make notes to be used for positive comments and post flight briefings.

...do not expect too much too soon!

Safety first officer

Discuss the role of the safety first officer. Both trainee and trainer should make maximum use of this crew member, who should be well briefed in his important role of monitoring the entire operation, and being ready to help you in any area. The safety first officer needs to be retained until you are satisfied that your trainee has a reasonable grasp of his duties and knows the limitation section of the operations manual and emergency drills.

Early pointers

During your discussion have your trainee make a note of some of the points that you want him to observe at the commencement of the training programme. Some of these could include:

- For each tour of duty, write out, or have available a list of flight numbers, flight times, check in times, pick-up times etc., if these have not been provided. These times will be used at the flight planning stage, inflight and on the ground when ordering taxis etc. This is a good beginning to flight management.

- Encourage early check in times at the start of the programme. Where possible, make use of the briefing rooms and equipment to outline the day's objectives. Avoid arriving at the aircraft without both parties having a definite goal in mind. Some suggested activities to be carried out during the early sign-on *buffer time* (say 30 minutes to 1 hour) include:

  - Expansion of flight planning methods and techniques
  - Practice full initial acceptance checks in daylight, using expanded check list
  - Short quiz sessions - limitations or emergency drills or the system that was set for home study
  - Practice good briefing, for take off, and instrument Approaches.

- Outline what you consider are essential duties at sign on time. These could include:

  - Arrive early, allow yourself time to avoid rushing, check the remainder of the crew, aircraft and paperwork
  - Obtain zero fuel weight - check for changes
  - Check aircraft ETA and gate number
  - Check fuel status - airports and operational notices
  - Check NOTAMS - notice board and memo book.

- Allow plenty of time for weather assessment, and flight planning. Encourage a systematic approach covering:

  - synoptic charts
  - forecasts
  - actual weather reports
  - note operational requirements
  - wind at all levels
  - sigmets
  - flight planning folder
  - notams
  - holding weather/traffic
  - load control.

- Aim to arrive at the aircraft at least 30 to 40 minutes, or on initial acceptance, ideally 45 minutes, before departure.

Encourage your trainee to have all charts and route data cards prepared for the route to be flown. Also, encourage him to crosscheck currency dates of charts and cards with other crew members. Stress the importance of order, sequence and a quiet methodical approach, with duties spaced out and correct priorities observed.

It's not the hours you have in, but what you have in the hours.

...practice good briefings!

Programme

Each training pilot will have an individual approach as to determining a programme of reading and instruction. A very broad guide, based on the experience of many trainers is:

- Have the trainee commence his operations manual study reading at flight planning, limitations and emergency drills

- Early in the training programme, just let him fly the aircraft to gain confidence and settle in. You assist and observe

- Commence a joint study on the normal operations section of the operating manual, to be completed approximately one third of the way through training

- Set your trainee homework. Study a system and discuss at the next flying session from the point of view of:

  - Cockpit switching and gauges
  - Limitations and emergency drills
  - Normal operation of the system
  - Non-normal operation.

Do not get involved in long, time wasting discussions on hypothetical or unlikely *what if* situations. Concentrate on what he can do about it in the cockpit. Emphasize the *need to know* concept.

- Take time to introduce your trainee to other staff members with whom he will come in contact. Show the location of phones, toilets, load control offices, notice boards, ATC briefing rooms etc. Give suggestions as to how to prioritize all required duties during short turnaround times

- Organize a progression of study through all relevant manuals and documents over the training period, refer to company syllabus

- Do not overload or over train. Fly the aircraft yourself, at least one sector in six. Be prepared to demonstrate in your flying and use *patter* (saying what you are doing as you do it) as necessary
- Keep in mind the need for your trainee to be proficient at support duties as well as manipulation. Give him plenty of practice

- Encourage self-criticism and open discussion. Where possible discuss problems at the end of a sector while recall is fresh, rather than wait until the end of the day

- Develop an awareness of the need for on time operation and economy. Do not set up instrument approaches that will cause the aircraft to run late. Remember that the instrument trainer and simulator can both be used to provide additional I/F work where necessary

- Keep en-route airborne conversation and discussion to a minimum. Monitor the flight path of the aircraft and your trainee closely

- Make use of all available time. Explain to your trainee that flying training is usually made up of at least three distinct stages:

  - The longest stage, of teaching, demonstration, application and repetition
  - A stage of gradual withdrawal of support
  - A stage of self-sufficiency in normal operation prior to check-out

- Be flexible but firm in your attitudes. Be prepared to use any method that works. Do not do it all for him, but do make yourself available.

Furthermore

As a trainer, you will go on learning yourself, for there is no better way to learn than to teach others. Do not be discouraged by a poor result, or a failure to pass a check-out. Try to obtain an accurate appraisal of what went wrong and why. Be honest with yourself and with written comment on training reports. Do not avoid making criticisms where necessary. Make sure that your criticisms are soundly based. If asked a question, to which you do not know the answer, say so, but attempt to research the problem and provide the answer. The question will probably arise again with a future trainee.

If you have problems with a personality clash, or friction develops in the relationship, discuss the situation with your trainee and with your flight manager. In rare situations, the solution is to pass the trainee to another trainer. Persisting with an uncomfortable or hostile cockpit environment is both dangerous and unproductive.

Familiarize your trainee with the flight manager's office and introduce him to flight operations department personnel. When telephoning your flight manager avoid discussing subjects that reflect adversely on your student without having discussed the matter with your trainee first. Be absolutely sure of all the facts of any contentious issue.

Discuss your problems, achievements and techniques with other training staff. Occasionally fly a supernumerary sector for yourself. To observe other crews in action can be beneficial. No one person has all the answers, no matter how experienced. If you find a good idea that works, share it with others.

...be flexible but firm!

Keep your goals clearly in view at all times. Fly the aeroplane to the best of your ability. Analyse problem areas and become a keen observer of the whole process of airline flying. As your trainees pass their check flights and go on to become operating crew members in the ranks of professional airline pilots, you will obtain a well earned sense of satisfaction and personal achievement.

Characteristics of trainees

- They are adults

Trainees are men and women who are usually mature intellectually, physically and emotionally. They generally possess above average intelligence and have demonstrated an aptitude for the task in hand.

- They are motivated

You do not usually have to interest your trainees in the subject. They want to learn. Trainees have varying expectations of the qualities they seek in their instructor, and their perceived difficulties of the learning task. They are achievers, used to passing examinations and achieving goals and so possess the self-discipline, which this requires. However, they also have a healthy respect for the problems of learning, particularly so amongst the older age groups. Usually the training programme is approached with some degree of apprehension ranging from mild concern to deep worry, depending on the individual. The question, w*ill I make it*? is ever present. Others have a very broad enthusiasm for the subject and seemingly boundless energy. You must be able to accurately assess all these qualities very early in the programme and adjust yourself to your trainee with appropriate, firm, fair but friendly guidelines.

- They have a serious purpose

They don't just happen to be there. It has taken a great deal of effort in most cases. They expect you to acknowledge this seriousness of purpose, as the outcome for them usually means, promotion and advancement in their career, together with the influence on pride, status, salary, working conditions, and self image, which the advancement represents.

- They are usually practical people

They want to apply the knowledge, theory and demonstrated skills, which you will supply. Generally, they are able to apply successful study methods and will want to know, *how? when? where? and why?* They will expect you to be practical, methodical, knowledgeable and keen, reflecting in some measure your interest towards them.

- Their reactions to you and the training will vary

They will appreciate that you know your subject and present it effectively. Conversely, they will quickly sense inadequate preparation, knowledge and skill. They will know whether you have a genuine interest in them and will quickly sum up your motivation. Reactions will not always be easy or predictable. Sometimes a reaction will be unrelated to the training and will simply represent what is often referred to as a *personality clash*, although because of mature attitudes, this situation is rare.

- They will come from varying backgrounds

They will have varying degrees of application, good general education, past experience, emotional stability and desire to achieve. Sometimes an inverse reaction will take place, often when you are younger than the trainee, you may initially be overawed by the experience, background and expertise of the trainee. If you apply correct principles and techniques, this situation is usually quickly overcome. Experience in flying hours is not always a good guide, as a pilot with fewer hours who has been operating in a demanding environment, may actually be more experienced than a pilot with a larger number of hours accumulated in a less demanding or less challenging flying environment.

...have a serious purpose!

Being a better training pilot

Flying training philosophy used to be based on two assumptions. Firstly, that only those with natural aptitude for flying would succeed and that most instructors could identify these *naturals* early on in flying training. Secondly, the assumption was that by subjecting trainees to a high degree of stress, this would help weed out those without this natural aptitude. Application of this philosophy, in an airline training environment, results in making flying training an unrewarding experience for trainees, who perceive every exercise as a potential chop ride, reducing the trainer's role to that of aptitude selector, not to mention having a negative effect on flight safety.

Perhaps understandably, there is a tendency to see *teaching*, rather than *learning*, to be fundamental in a training pilot's role. This leads to an emphasis on syllabus content in flight instruction, placing liability on the trainee to observe and reproduce skills. It should be recognized that more weighting needs to be given to the central role of the trainee, rather than that of the instructor.

With the increasing complexity of modern aircraft, the contribution of psychological factors to efficient operations is now widely accepted. Increasing attention being given to training in aviation resource management points to a changing situation. This hopefully will lead to greater attention being paid to the psychological factors associated with skill acquisition.

...practice makes perfect!

In the late 1980s, the British Royal Air Force examined the reasons for the suspension of trainees from flying training duties. Their conclusions were that the majority of failures resulted from insufficient spare mental capacity when subjected to high workload, a lack of awareness, slow thinking and poor retention of instruction. A slightly different bias needs to be applied to airline trainees, which should address the psychological issues in skill acquisition, namely, motivation, stress and anxiety.

Research has demonstrated that individual differences in learning rates do vary considerably. Not everyone makes the same progress. Moreover, the rate of learning in the early stages is not necessarily a good predictor of ultimate levels of performance. Quite often, within a fixed length course, extra time spent consolidating the basics in the early stages may pay dividends later.

In skill acquisition, it is only true that practice makes perfect where the trainee has grasped the fundamentals and is able to place new information into an appropriate context. When a skill becomes automatic and well established, it is very resilient. It leaves the trainee more receptive to dealing with peripheral activities, which are the lifeblood of a good airline pilot.

Why trainee pilots fail

- Workload and mental capacity   66%
- Awareness   48%
- Slow thinking   36%
- Retention   35%
- Performance under stress   17%
- Consistency   16%
- Decision making   15%
- Accuracy   15%
- Progress   12%
- Overconfidence   11%

Stress and its effects

How trainees *feel* about their trainer/instructor is important. One of the biggest areas of criticism, from trainees, is the often-perceived poor standardization from their training pilots. There are still vague areas, where as a trainee, you are expected to do a certain thing, or approach a certain task, in different ways depending on which trainer you have. This causes a great deal of anxiety and frustration and is a great de-motivator. Most people accept that flying *can* be a stressful activity. For the trainee there are many additional stressors created by fear of failure, anxiety, keeping up with their colleagues and coping with the personality of their trainer/instructor. All of which can add a very negative element to the learning process.

Some trainers/instructors, far from helping reduce this stress, actually are responsible for increasing it. A survey conducted in the early 1980s found that the most stress-producing agent in pilot training is the instructor himself. The *positive* trainer/instructors, using techniques of encouragement, acceptance, praise and generally perceived good instructional techniques, produces less stress on the trainee and results in overall better performance. On the other hand, the *negative* trainer/instructors, relying on heavy criticism and harsher methods significantly raise stress levels in trainees with a significant stifling of performance.

What is stress?

Stress could be defined as the way your body responds to the demands of your life style. Stress is not just negative; it is an important part of increasing responses, improving speed, strength and ability to work more effectively. Stress is with us all the time. Problems arise when stress levels are not managed correctly, either by trainees or more likely by overzealous instructors.

All too often, some trainer/instructors fail to understand and recognize the symptoms of negative stress. It should be appreciated that a culture that boosts the trainees' self esteem is more effective in improving their performance, than one where they are being continually being put down, punished and scared into response.

Stress management

The key to stress management is awareness.

- How does stress affect you?
- Which stressors affect you in particular?
- How can you become more aware of your stress response?
- What is the strength and duration of your stress response?
- Which stress management techniques suit you?

Positive effects of stress

- Increases alertness
- Improves vision
- Strengthens ability and resolve

- Mobilizes resources
- Stimulates
- Adds zest to life.

Negative effects of stress

- Poorer health
- Lower energy
- Decreased performance
- Unsuitable behaviour
- Increased forgetfulness
- Inflexibility.

Examples of stressors

Physical	*Social*
Heat	Training
Cold	Competition
Noise	Power struggles
Violence	Interruptions
Illness	Career
Traffic	Lifestyle
Poor working environment	Muddled communication

Whether we like it or not, whether we recognize it or not, we are constantly being subjected to stress. We all have consciously, or unconsciously, built-in defence mechanisms to cope with everyday stress. Being aware of what stresses you most and what your particular stressors are, will give you, either as a trainer or trainee, a clearer picture in how to take avoiding action. There are stressors that you can do something about, there are stressors you can get help to do something about, and there are stressors you will have to learn to live with.

To be effective, trainer/instructor pilots need to be aware of these negative qualities. By applying sound psychological techniques, or just pure common sense, their training effectiveness will be improved, making the task a more rewarding experience for not only the trainees but for themselves as well.

Creating a sound basis for good instruction is the training pilot's aim. The next chapter outlines some of the basic requirements.

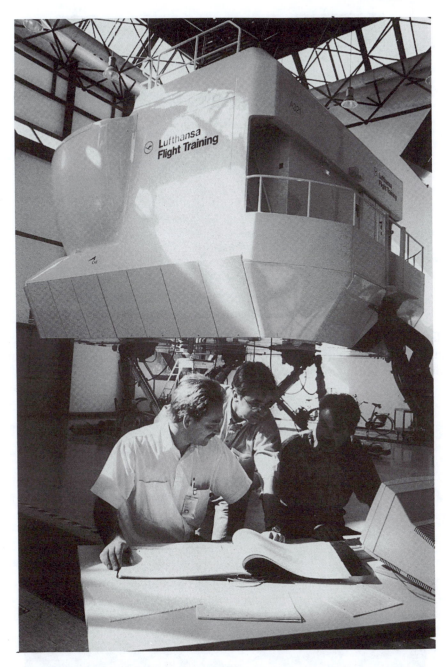

Simulators need to be maintained in peak condition if they are to offer realistic training value.

Flight simulators, like the B777 and A340 depicted here, can reproduce nearly all the critical flying conditions for crews completing type conversion and proficiency checks.

3 The basis for good instruction

A list of the specific duty and skill areas required of a training pilot would read something like this:

- Conduct good briefings
- Impart specialist knowledge of applied techniques
- Assist with research of available reference material and direct efficient study methods
- Conduct guided discussions
- Conduct demonstrations and talk through (patter) in aircraft manipulation
- Ability to ask questions, review and summarize
- Conduct soundly based critiques and evaluation
- Competent processing of training documentation
- Ability to write concise, accurate reports
- Maintenance of sound personal standards of aircraft manipulation, procedural knowledge and airmanship.

Your general background and training will of course help you in most of these areas. It is in the specific area of how to go about simple basic instruction, which causes the most concern. Can I be effective? Can I *get the message across?* Can I see what he is doing wrong, and do something effective about it? Fortunately, the *teaching content* of the job is fairly low key and informal. None the less it is most important. Well done it is satisfying, done haphazardly it is frustrating.

Most people like to feel that they are being of positive assistance to a trainee pilot whether they are gifted with any instructional skills or not. These skills are simple to learn and implement. Sophisticated methods are not required. Enthusiasm goes a long way in any field or endeavour. The single key word for success is *preparation.* For our purposes we shall look at three key areas of the teaching process:

- Briefing
- Discussion
- Questions and evaluation.

Conduct a good briefing

During the course of a training programme you will usually carry out a number of instructional sessions with your student in order to improve his knowledge in some specific area.

These teaching situations tend to occur in the crew briefing rooms initially, with further back up instruction in the aircraft cockpit and overnight hotel rooms. Opportunity for briefing also exists in the technical training school and the simulator complex, particularly in association with the cockpit systems trainer.

...conduct a good briefing!

Instruction carried out in this manner can be very effective using nothing more than a white board, some quick sketches and a few well-chosen words. It sounds easy, it looks easy, it is easy, but how often is it poorly done!

The steps

Let us look at some of the steps involved in order to get good results. In any teaching situation, there are four basic steps:

- Preparation
- Presentation
- Application
- Review.

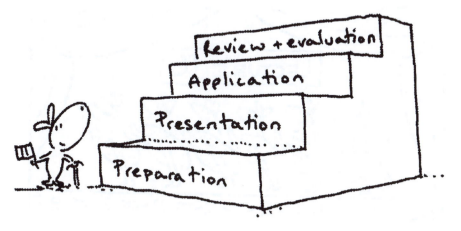

...teaching steps!

Preparation

Research your subject. Make written notes, short statements, memory joggers, and reminders of any useful anecdotes. Cover the ground, but keep it brief, can you use a mnemonic? What sketches will you use? Practice drawing them, both on paper and the white board. Try to be quick and neat.

Do not use too many labels on your drawings. Remember the usefulness of the *profile-plan-profile* diagram for illustrating circuit flying. Do you have pens and duster? Carry your own if necessary. Using a white marker board and pens, blue, red, black and yellow are effective colours. Make sure of your facts and numbers and commit as many as possible to memory, to avoid having to refer to your notes. Rehearse the words and phrases you intend to use. Do a *dry run*, recording your voice if possible. This is usually a revealing and useful experience.

Pay attention to your choice of words, enunciation and articulation. In any form of speaking before an audience, confidence is everything, even in a *one to one* situation such as a simple briefing. Preparation gives confidence. Confidence gives fluency, polish and credibility. Both parties will enjoy the briefing when it is well done.

...presentation and application!

Arrive early and check that the room is available and clean. Make sure that the white board is clean, as a dirty board is very distracting and lessens the impact of your sketches. Check your equipment, pens, duster, and any teaching aids you intend to use. Turn the lights on. Check that there are sufficient seats available. Do you intend to use a wall map and pointer? Is there spare notepaper available? Open the door or window to let in some fresh air.

Give a brief introduction to your subject. State your aims and objectives. Outline methods and techniques involved. Speak clearly and confidently.

When using the board for sketching, stand side on. Do not obstruct your trainee's view. Try to develop your sketch as you talk. Avoid too much writing. Remember that it is a briefing. Make all your diagrams as large as the board permits.

Allow approximately two minutes for your introduction, ten minutes for your presentation, and three minutes for summarizing, making fifteen minutes in all. Have an awareness of what is horizontal and what is vertical on the board. If you are not particularly artistic, keep your sketches to simple diagrams and plan forms. They will be no less effective. Tell your story completely then stop.

Review

Go through the main points of the subject once more, using a fluent explanation supported by your white board diagram. Some subjects lend themselves to a summary treatment using a mnemonic, or similar form of abbreviation.

Allow a short time for some simple direct questions from you. After the subject has had further practical treatment in the form of a flying session, your trainee may have further questions to ask. Prior to another flying session, which will include the subject under discussion, use a further oral quiz to review and consolidate.

...confidence!

Briefing subjects

Subjects that are suitable for briefing treatment could include the following:

- Take off and clean up phase
- Cross wind take off
- Early turn after take off
- Windshear

- Special airport terrain and procedures
- Performance knowledge
- Instrument scan
- Phase 1 (recall) drills
- Circuits and landings
- The descent profile
- Final 30 miles
- ILS profiles
- VOR / DME arrivals
- NDB approach
- Bad weather circuit
- Turbulence penetration
- Simplified systems.

Use a briefing to clarify and amplify any problem areas, which occur during line training.

Discussion

You will spend a great deal of the time with your trainee in discussing various aspects of the line training programme. If the discussion is planned and guided, very effective learning will take place. The aircraft operating manual is the prime reference. Do not attempt too much at any one session. Endeavour to correlate the material from the various reference sources.

Example

A suggestion, for studying and discussing the hydraulic system, might be as follows:

- Talk about the general layout and function of the system (reference engineering notes)
- Learn all relevant limitations (reference aircraft operations manual)
- Find the appropriate switches, gauges and circuit breakers in the cockpit
- Make a list of all hydraulically operated components
- Point out the location of components, reservoirs, etc., on the aircraft
- Discuss and list the non normal operation of the system (reference aircraft operating manual)

- Discuss the emergency operation of the system, or components, and systems available on emergency power
- Use anecdotes which will relate the hydraulic system in any incident or accident, about which you have facts available
- Get your trainee to tell you *in his own words* how the system works
- Conduct oral quiz and question sessions from time to time as a refresher.

Try to *package* your information in this way. Most systems can be covered in a similar manner. Very few teaching aids are needed, but do not forget the value of a simple sketch. Much good information has been passed on using a quickly drawn sketch on the back of the flight plan. Organized thinking and preparation is the key to good, guided discussion. Start from the simple and work towards the more complex. Do not allow your discussions to become haphazard or open ended, with too many unanswered questions. Avoid long rambling explanations. Do not be sidetracked. If you do not know the answer, say so, but find out and discuss it further.

Listen to your trainee, and let him relate the subject to his previous experience. Encourage an enquiring approach. Keep your discussions reasonably brief and always to the point. At the end of each session always nominate the next subject for study discussion. A study/discussion session can be enjoyable, constructive and informative.

...encourage an enquiring approach!

Questions

The effective use of questions in teaching is a highly developed skill. Endeavour to keep them simple in structure, meaningful and relevant. In terms of their characteristics, questions are classified as follows:

- Leading
- Rhetorical
- Direct
- Open/closed
- Reverse
- Rebound
- Follow up.

Good questions, well thought out, should contain the following elements:

- Have a specific purpose
- Be clear in meaning
- Contain a single idea
- Stimulate thought
- Have only one correct answer
- Relate to previously taught or discussed information.

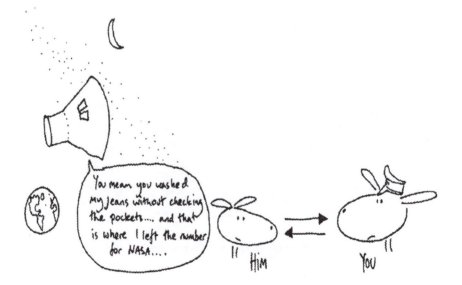

...stimulate thought!

In order that a better understanding of good questioning practice may be obtained, let us look at the various types of questions in more detail. Generally speaking, the problem is related to anticipating areas of the subject matter that will require further thought stimulation. Framing good questions will achieve this result.

- Leading questions

An opening question, usually directed at a group to help get a guided discussion started. An answer is not required immediately. A well-chosen question often has the effect of silencing a group, and stimulating their thoughts, thus setting a mood.

> *e.g. Let us discuss an en-route engine shut down. What are some of the flight management considerations?*

- Rhetorical questions

Similar to a leading question in that it is used to stimulate thought or build up expectancy prior to a discussion. You always answer it.

> *e.g. What are we going to do about it? pause ... I'll tell you what we are going to do about it ...... (explanation).*

...a leading question!

You answer the question yourself, providing the answer *you* want, to lead the discussion in a particular direction. Rhetorical questions lose their impact when used too often e.g. *in some political speeches,* and are generally more suitable for a group situation.

- Direct questions

The most common type of question. An answer is required promptly. Can be used in an individual or group situation. A direct question to a group is effective when followed by a short pause before being directed to an individual, thus keeping the whole group concentrating; *is he going to ask me?* Random questions are more effective than the *round robin* type, provided that the distribution amongst the group is equitable.

Direct questions are derived from the considerations of:

How? How many? When? Where? Why?

- Open and closed questions

These can be broadly classified into five main types of question according to the answer or response they generate.

Factual answer	Questions on aircraft performance or limitations. These questions are effective when used in oral quiz sessions. Short answers, numbers, facts, but avoid *yes* or *no*
Sequential answer	Questions on EMC drills and memory lists
Descriptive answer	How?, When?, Where? *Describe the A320 descent profile?*
Explanatory answer	The trainee explains the answer in his own words. Questions on aircraft systems or airways operating procedures
Reasoned answer	The *Why?* answer, usually to the most difficult questions. The student must draw on knowledge, memory, reasoning ability and be able to interrelate and express ideas. For example,

questions related to specific system inter-relationships, such as how does this electrical switching affect the hydraulic system, or emergency system operation? Questions will be generally more complex or philosophical in nature. Answers will indicate not only an understanding of the subject, but a depth of reading and interest as well.

- Reverse questions

You may use a reverse question in answer to a trainee's question, in order to stimulate more thought or re-direct his thinking. Do not overdo this technique as it can become very irritating.

- Rebound question

Used in a group situation. If one trainee does not know the answer pass the question on to the next one. Alternatively, a question asked of a specific individual may be re-directed to the group.

- Follow up question

After a discussion develops, you may use a follow up question to re-direct discussion back to the leading or introductory question. A follow up question may be a rephrased question seeking a better answer to an earlier question.

Types of questions to avoid

Asking *do you understand?* or *have you any questions?* have no place in effective quizzing. Assurance by your trainee that he does understand, or that he has no questions, provides no evidence of his comprehension, or that he is even specifically aware of the subject being discussed. Remember that, at least initially, you have to ask the question to test for comprehension of the subject.

Do not let questions become time wasting, either through poorly framed questions on your part, or a trainee who seems to ask questions for the sake of it. A question must seek information that is needed. Some individuals, particularly at meetings or group discussions, ask questions out of a pseudo-intellectual approach.

Avoid using the following types of question:

- The puzzle

A question so badly framed and *long-winded* in presentation, that the trainee is confused by the question and distracted. This produces a very negative and frustrating situation.

- The over-size

A question that is not sufficiently specific. *What is the most important thing to remember prior to take-off?* There are so many important things to remember prior to take-off, that the answer will be a reasoned guess.

- The toss-up

A badly framed question with two or more possible answers. The student will eventually guess the one he thinks most likely.

- Catch questions

These are most unproductive, and give your student the feeling that he is involved in a battle of wits with you, rather than a genuine pursuance of knowledge.

- Over simple and over complex

These are unproductive and time wasting questions.

- Irrelevant questions

These are time wasting and tend to make you look foolish.

- Closed questions

Producing yes or *no* answers:
One word answers may well be the product of a good guess and nothing more. e.g. *do you know the single engine climb profile?*

Finally, never fall into the trap of asking a question to which you do not know the answer.

...make you look foolish!

Fault analysis and critique

In the flying training process, fault analysis is necessary at all levels of training. The ability to criticize effectively does more to separate the successful training pilot from the poor one, than does above average flying. The sole purpose of fault analysis is to improve the student's future performance. A valid critique contains three essential elements:

- *Strengths*
- *Weaknesses*
- *Specific suggestions for improvement.*

Without each of these elements, fault analysis is invalid, as it does not accomplish its purpose.

Strengths are analysed to fulfil two purposes. The acknowledgement of the correct procedure reinforces learning, and it encourages the trainee, satisfying a psychosocial need while gaining his attention. These are mandatory in improving future performance. Weaknesses should be noted in such a way that the trainee is clear on the precise points involved and reasons for the poor performance. Constructive suggestions should follow immediately.

One of the major errors that can occur during fault analysis is when you, in an attempt to be accurate, immediately emphasise all the errors that were committed by your trainee during a sequence. This is a completely negative approach and does not give him the necessary reinforcement to make the experience offer any satisfaction. A much better method, which also contributes to the principle of effect, is to firstly point out the positive aspects of his performance and then discuss the errors, which were committed. Whatever the learning situation, it should contain elements that affect him positively and give him some feeling of satisfaction. Each learning experience will not be entirely successful, nor does he have to master each lesson completely. However, his chance of success will be increased if there is some sense of accomplishment and the learning experience is pleasant.

The principle of feedback

All trainees need to know both what they are doing correctly and what they are doing incorrectly. In debriefing, it is easy to concentrate exclusively on those aspects of the trip that need attention. If the good aspects of the trip are ignored, your trainee leaves the briefing with an unbalanced view of his overall performance. In airborne exercises, you need to set demonstrations and standards that your trainee understands.

Before fault analysis begins, give your trainee credit for those items that he has performed correctly. Any errors that he has made can then be pointed out, followed by a positive statement on how such errors can be corrected or avoided during future practice.

Do not expect perfection after one attempt. Identify and correct the major errors one at a time and work down to the smaller ones in turn. Give him an opportunity to analyse his errors and correct them, before you offer guidance. In making your assessment of his ability, determine first whether he had made any progress or whether he is consistently making the same mistake.

If errors are persistent, ask yourself:

- *Did you cover too much?*
- *Did you ask him to observe too many things at the same time?*
- *Is he hearing you clearly?*

- *Does he really understand you?*

Encourage your trainee to give a constructive analysis of his performance at the end of the flight, and analyse faults in the following order:

- *Symptoms* - were they recognized?
- *Cause* - is this understood?
- *Result* - what would this have led to?
- *Correction* - what technique should be used?
- *Prevention* - for future practice.

The next chapter investigates the main key to all good instruction, that of communication.

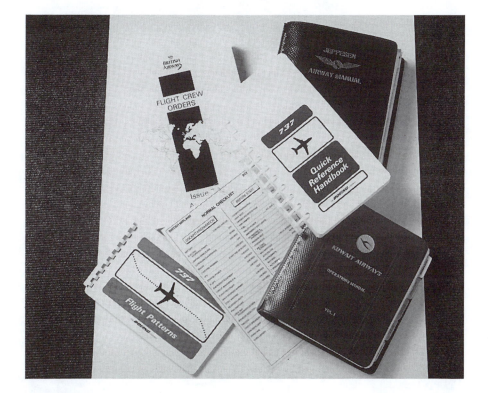

Tools of the trade.

Basis for good instruction.

4 Communication

A training pilot is both instructor and communicator. The role involves some teaching concepts, *stating facts, making assertions,* and some counselling concepts, *listening to your trainee, and diagnosing problems.* This chapter looks at the process of communication from both these aspects, and examines some of the basic elements and techniques involved.

Effective communication

It is as well to remember that we are dealing with complex areas of human relationships, especially in the counselling area. All discussions with your trainee should be completed in a suitable atmosphere free of distractions. The trainer will need to approach the subject of communication with a large measure of patience, tolerance and sensitivity.

The communication process

Communication takes place when one person transmits ideas or feelings to another person. Its effectiveness is measured by the similarity between the idea transmitted and the idea received.

In human communications, some form of suitable encoding and decoding is always taking place. The three elements that follow are dynamically inter-related and *reciprocal.* The factors at work are complex but what must be understood is that: Communication is a *two* way process.

Attitude projections and judgemental response

When two people talk, or when one talks and the other listens, there are some abstract and subconscious messages being sent, received and assessed; these are part of the encoding and decoding process mentioned earlier.

The basic principle is simple:

- The better the quality of the transmission, the better the message can be received.

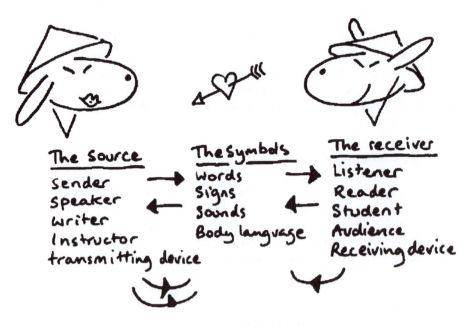

The Source | The Symbols | The receiver
Sender → Words → Listener
Speaker ← Signs ← Reader
Writer ← Sounds ← Student
Instructor | Body language | Audience
transmitting device | | Receiving device

...encoding and decoding!

When the trainer is making statements as assertions, both he and his information are under assessment and judgement by the receiver, *the trainee*. It therefore becomes a case of trying to transmit as many positives as possible. He must look the part, be convincing, and his information should be able to stand up to close scrutiny.

As in so many other aspects of life, when it is necessary to create a good impression, the following factors should be *all positive*.

Attitude projections

• Appearance	Clean, neat, groomed, professional
• Speech	Pleasant, authoritative, clear, well modulated
• Body language	Suitable facial expressions, movements, proximity, gestures.

- Enthusiasm   Must appear to be enjoying the job, with warmth, friendliness and desire to help
- Animation   The intangible area which makes such a strong impression, includes sense of humour, wit, presence, flair, empathy.

The whole approach from a personal point of view is designed to show that the trainer is:

- Positive and confident
- Believes in what he is doing
- Has experience in the field
- Has knowledge and understanding if his subject
- Is able to stimulate and inspire suitable emotional responses.

A tall order perhaps, but can be done well within the limits of most personalities.

... effective communication!

Judgemental responses

It is important to remember that in basic human communication when information is being transmitted, the first three areas of judgement on the part of the receiver are usually:

- Like or dislike
- Respect
- Credibility before any assessment of the actual material is made.

Other factors of importance in communications are:

- Accuracy of data
- Avoidance of repetition or qualification
- A strong command of vocabulary
- Good diction and articulation
- Material within the experience range of the receiver
- Variety and flexibility of communication symbols.

Communication between individuals

Let us look at a random selection of human communication situations. You will be able to think of many others; try to think of good communication *situations* in which you have been involved.

Conversation: Universal, deceptively simple, infinite variety. Always enhanced by good thought processes. Watch conversations being conducted in different languages and note the body language.

Telephone, immediately less effective than face to face, we lose some of the symbols and rely more on pauses and silences. Some people are not comfortable with telephone conversations.

...able to stimulate and inspire suitable emotional responses!

Audience involvement

Debate/Discussion/Argument/Lecture/Conference

Thought processes can be more directed, controlled and disciplined than conversation. We rely on additional symbols and body language. Next time you are at a meeting or a conference, study the transmitters and the receivers; note the symbols and body language.

Communication with limited access to the receiver

Television presenter/film actor

Very controlled, more difficult, no obvious audience but still using the same symbols. The receiver relies very heavily on the face in general and the eyes in particular. Voice is used to greater effect.

Observe the eyes, and method of delivery of the successful TV presenter. Can you feel that he/she is speaking to you? Do you feel a response? Is communication taking place? Does the voice *get through* to you?

Barrister, summing up for the jury

The barrister never actually engages the jury in conversation and yet he communicates, sometimes at a very intense level. Throughout the trial he has studied the body language of the jury, *and they his*. Note the use of rhetorical questions, posturing, movement, distancing, stance etc. He has their attention, and concentration is taking place on both sides.

Foreign language difficulties

Many substitute techniques used:

- International airport symbols
- International traffic signs
- Numbers can replace words
- Money systems; it may take time to learn the language but exchange notes and currency systems are quickly learned
- Much more reliance on sign language, body language and gesturing.

Technical and specialized communication

Written and coded communications are outside the scope of our interest except to say that a de-personalizing effect is noticeable with, say, telex, telegram, metfax, and other coded data.

It is worth noting here that the *Check List* that we use in our aircraft is a *clipped* method of communication, which can be made self-correcting. The check list is in the form of *challenge and response*. After the challenge is called, a response is given. The response must be correct for the check to proceed.

Technical jargon/abbreviations/computer language/multi-yuck-speak

Highly specialized communication has its place in aviation, where a real need for clear, precise and unambiguous communication minimizes mis-understanding. However, it should be appreciated that under some circumstances it is not *always* as effective as plain language. Reliance on too much technical jargon in any field may actually impede communication.

The other characteristic, which is of interest to us, is the need that humans have, to *humanize* their communications, and revert to plain language.

In the early days of space exploration, elaborate codes were devised and a great deal of attention paid to phonetics and the avoidance of ambiguity. Rigid radio telephony (R/T) procedures were set up to avoid the possibility of confusion. However, the need to *humanize* was so powerful that in a very short time, plain language, using buzzwords was substituted. Astronauts joked and exchanged seemingly trivial banter from deep space, to their fellow humans at mission control on the ground. A similar humanizing tendency is observable in aircraft R/T communications procedures.

Tactile communication

Some form of touch or hands-on approach is used as a substitute for other symbols:

- Sign language and Braille
- Learning the piano with a teacher
- Learning to ski with an instructor
- Learning to fly a light aircraft.

A tactile approach is re-emerging as a technique for communication in high technology situations involving computer keyboards and switching, and of course is very relevant during hands on simulator training.

Measuring the success of communication

From a study of the foregoing examples, it can be seen that feeling and emotion are always present in communication. It is largely through these elements that we can judge the degree of success at the attempt at communication. We have so far looked at communication largely from the point of view of the sender. Remember that communication is two-way, that we are both sender and receiver and may have to vary our approach to sending, on the assessment of the messages we are receiving.

Verbal and language skills

Verbal and language skills are the next area of study, but before leaving the exposition side of communication, let us remember that communication can be said to succeed as a process when:

- The reaction and response of the receiver is the one intended
- The idea transmitted is the idea received
- The receiver's understanding of transmitted concepts bring about an appropriate behavioural change.

...a tactile approach!

Communication skills

Communication is a two-way process which, greatly simplified, can be said to be a human signal, sending and receiving. Most people are able to send information with adequate skill, but very few are good receivers. For this reason, many attempts at communication, from open or public interaction, to the intimacy of married life, do not succeed. Whilst the receiver may hear the other person, he is not necessarily listening with adequate understanding for either the subject matter or feelings being expressed. His responses will therefore be inadequate and inappropriate to the sender. Ineffective communication will result in frustration.

In this section, we attempt to look at some of the techniques, which a trainer may employ, to better communicate with his trainee. In particular, in the areas of listening and verbalizing feedback, trainers are urged to widen their reading in communication skills generally, as greater effectiveness will flow from better human interaction in any endeavour. Listening and counselling skills can also dealt with, in educational courses, provided by companies from time to time: *human factors, aviation resource management courses.*

Being a good listener is a skill, which can be learnt and developed, and from the trainer's point of view, is the stepping stone to being an effective instructor. In the cockpit environment, during a line flight, the priorities are to get the aircraft safely and efficiently from departure point to destination. Teaching, talking and listening opportunities are few. There is a good deal which the trainer may learn in the cockpit environment from voice modulation, and observance of body language, but generally the opportunity to talk, discuss problems and listen, will arise away from the aircraft. All trainees want to talk and they want you to listen. The first step is to provide a suitable listening environment and then to concentrate on receiving and interpreting the message which your student will send. You have to be ready to listen and to show that you are ready.

Empathy

The quality of empathy is a prerequisite of a good listening situation. Empathy is being able to feel what the other person is expressing. A captain *may* have empathy with a first officer, because he himself may have spent ten years in the right hand seat. The very nature of the complex human personality has a profound effect on whether empathy is present in these situations. Your trainee will make certain assumptions about you, your experience and ability, but he will feel empathy as a tangible quality. Lack of it produces disappointment and triggers negative reactions.

Empathy cannot be faked; it is there in the tone of voice, the movement, proximity, body language and facial expression. It is not sympathy but implies a willingness to listen and understand. Faced with a problem we may say that the difference between *apathy, empathy* and *sympathy* would be as follows:

- Apathy   -   *That's your problem*
- Empathy -   *Sounds as if you're really concerned about this*
- Sympathy -   *How dreadful, I do feel sorry.*

Warmth and sincerity of purpose are the two qualities most required.

...apathy, empathy, sympathy!

Attending skills

After flying, or simulator training, is often a good time to talk, but *not* always. The responsibility for discussing training issues is a shared one. The trainer should involve the trainee about when and where they should talk things over. In the structure of your time with your trainee, try to gauge a suitable time and place when you can discuss problems. Do not intrude when a period of rest or private consolidation may be more appropriate for your trainee. Hotel rooms with comfortable chairs, good lighting and quietness are usually the most suitable. Avoid having a desk between you, it is a surprisingly powerful block to free communication, as it implies authority.

Attending is giving your physical attention to another person, whilst listening. Set yourself at a comfortable distance from the speaker and adopt a relaxed style with all open body posture. Face the person, and if seated lean forward slightly. Make reasonable eye contact, not staring, but show that you are interested. Concentrate. Try very hard to listen to both what he is saying and gauge his feelings. If a thought is triggered off in your mind, avoid cutting off his line of thought with yours *by interrupting* but rather save your thought for a lead on the subject later.

Allow pauses and silences now and then to give the speaker *space* and help him feel relaxed. Every now and then, an acknowledgement may be appropriate, using body language, a nod, a gesture or smile. Use some words to respond initially. *Yes* or *I see*. These are called *minimum verbal encouragers*. They do not start a new line of thought. Good attending skills can work wonders in any person to person, or people to people environment. When a good speaker has an audience enthralled, *he has them on the edge of their seats*. They are not only leaning forward but sitting forward; conversely, folded arms means defiance; wrong body posture may mean disinterest, *giving the cold shoulder* and so on. Try to avoid sitting on the edge of the desk or standing. This would tend to show that you are the superior party.

We all know that it is an impressive experience to talk to a person who is there for you only, that you have his total interest and concern at that moment. In professional relationships, these inter-personal moments are relatively rare. A doctor may have it with his patient, *sometimes* or a stockbroker giving a financial summary to a client. A training pilot should strive to make good use of time with his student so that relaxed and meaningful communication is possible.

Summarizing attending skills

- Face the other person *squarely*
- Use an *open* posture
- Tend to *lean* slightly towards the other person
- Establish a comfortable *distance* from the other person
- Establish appropriate *eye* contact
- Try to feel *relaxed* even though you are concentrating.

Being a good listener

One of the skills of being a good listener is to *give the other person room* or *stay out of the way,* so that the speaker can have good uninterrupted thought processes and so produce a good quality message. Listeners tend to make two common mistakes, either interrupting too much and making statements which tend to deflect the speaker, or mistaking complete silence as the best listening environment.

A good listener needs to develop the use of what are called *following skills*. These include the use of:

- Minimum verbal encouragers
- Door openers
- Open questions
- Attentive silence.

Early stages

The speaker may not know how to start. His body language and expression may show a problem. He may even be expecting to be criticised after a day's flying when things did not go so well. A door opener could be, *you didn't seem happy with your descent profile today, care to talk about it?*

An open question may be appropriate. However questions should be used very carefully especially in the early stages of a dialogue, *how did you feel about today's flying?, what's on your mind Alan?, how did you feel about your landings today?*

It is most important in the early stages of a conversation for the listener not to try to solve any problems, offer any solutions, or instant cures. Just get the speaker relaxed and speaking. Allow silences and time for him to collect his thoughts. Use some minimal verbal encourages, *good; right; that sounds good....* The listener may be tempted with an overpowering urge to offer solutions and ask questions, but indirectly these are usually counter productive, and are called roadblocks to communication. Initially the speaker may be inclined to be defensive of a particular performance and offer some excuses or *smokescreens*.

He may tend to blame the weather, air traffic control, or even imaginary shortcomings in the aircraft for some problem, which he is having. This technique on the part of the speaker is called *shifting the blame*. It usually disappears when he determines that he is secure with the listener, and that he is being given a fair and honest appraisal. The listener has to be patient.

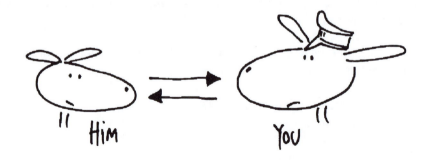

...shifting the blame!

Attentive silence

Learning the art of a silent responsiveness is essential to good listening. Most listeners talk too much. Silence on the part of the listener gives the speaker time to think what he is going to say, and to proceed at a comfortable pace. During silences, both speaker and listener will be communicating with body language and the listener will be making use of sound attentive skills, as well as considering a variety of responses he might make at the appropriate time. A skilled listener will also space the silences with *followers* or *encouragers* without changing the direction of the subject matter.

Reflective listening

When a speaker is under stress or has a problem, reflective listening is a good skill to use. The listener provides a mirror for the speaker, paraphrasing his statements and concentrating the focus. Reflective listening seems unnatural at first but is a sound technique when not overused. Reflective listening shows a willingness to try to understand and help the speaker to explore and clarify an unclear situation.

Speaker: *I just can't seem to land the aircraft consistently well. I'm doing something wrong.*

Listener: *So, you're having some problems with the landing?*

Speaker:	Encouraged to go on: *Yes, I have the speed right, and I feel the power is about right, but just seem to lose it in the last couple of seconds.*
Listener:	*The speed is right - the power is right...*
Speaker:	*Do you think it would be...*

Although it may seem odd to reflect the statement in this way it does allow the other person to go on speaking, to direct the flow of thought and perhaps lead to his own problem solving, which is a highly desirable objective. It is also a thought spacing process, which aids concentration and visualization.

Consider another example:

Speaker:	*This auto throttle system is annoying. I don't like it.*
Listener:	*You feel that the auto throttle system is not helping you.*

Notice that there is no attempt to give a closed response e.g. *It really is a good system you know - it works well.* This response would produce frustration in the speaker. There should also be no attempt at problem solving at this stage. Your aim is to get him to speak and find out himself why he is having trouble with the system.

Reflective listening is a means of giving *feedback* to the speaker in three main ways.

- Paraphrasing: Re-state the message, using the speaker's own language. This lets him know that you are listening and encourages him to go on
- Summarizing: Capture the essence of what the other person is saying in fewer words, a quick word picture from you the listener
- Drawing implications: Put into words a message which is implied but not actually stated.

Open or closed responses

An open response from the listener is also an encourager and shows that the listener is accepting what the speaker feels as well as what he is saying.

- Speaker: *I passed my simulator training session. It wasn't as bad as I thought it was going to be.*

- Listener: *You're pleased with your performance in the simulator ...*

- Speaker: *Yes - remember that I was having trouble with ...*

Closed response

This is a terminator of conversation, the end of a discussion and perhaps the classic *put down.*

- Speaker: *I passed my simulator training session. It wasn't as bad as I thought it was going to be.*

- Listener: *Oh - the simulator is nothing, most people don't have any problems.*

The listener is apparently unwilling to try to understand the speaker's feelings and tries to close off the subject. An important point to note is that an open response is not praise but rather an encourager and implies a degree of mutual respect. A closed response is not condemnation but rather disinterest.

Open responses are generally better in early problem solving and a good general response is: *Show me how you think it works, I'll follow you through, and we'll work it out together.*

Conversational skills

After an initial period of listening where the subject matter may be generalised, the listener can often sharpen the focus with some well chosen questions, usually very short, so that the speaker is encouraged to go on speaking.

...being a good listener!

Asking for specifics

The listener may invite a problem centred response by asking *what?* rather than *why?*

- Specifics

  What happened? - focus on the situation
  What did you do? - focus on the actions
  How do you feel about it? - focus on emotive reaction.

- Value judgements

  Is what you're doing making things better or worse? -
  consider the results or outcomes
  Is what you are doing, helping you? - consider the feelings
  about competence/performance of a skill.

- Consequences

  What would happen if..? - results
  What is likely to happen?- understanding the consequences
  What is the worst possible thing which could happen ? -
  setting the limitations.

- Focus on plans and actions

  What can you do about it? - available resources
  What are you going to do about it? - decision making

- Follow up and consolidation

  How is it working?

  Can you give me an example of the problem?

  In what way is it not working?

  Can you be more specific?

Problem solving is much more effective when the student is led into his own solutions and made to see both the reasons for, and consequence of, the actions he decides upon.

Verbal and language skills in conversation

Just as there are not many good *listeners* in society, there are also relatively few people who are good conversationalists. Good quality conversation is essential in the *one to one* teaching situation with which we are dealing. Good conversation is free flowing, interesting and stimulating to the participants. Patience, courtesy and consideration are necessary.

We have all had experience of *difficult* conversations. The frustration of awkward responses; no response; being *cut off* in mid sentence; *not being able to get a word in,* and so on. Sometimes we become impatient with rambling, incoherent thought processes, the feeling that the other person is not listening, or the total rejection of our point of view. Conversation is dependent on our mood, energy, circumstance, sense of priority and sense of security. Our thought processes range from *boring party small talk* with a dull person to heated and animated discussion on subjects about which we feel emotionally involved. The trainer should endeavour to exercise at least a small amount of *control* in conversation with his trainee, but this is not easy and effective listening as previously discussed may be a better alternative. We are using words as our symbols of communication strongly reinforced by body language, movement, animation, varying tone, pitch and volume in our voices.

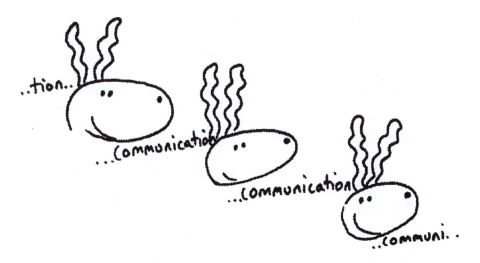

In an effort to keep the conversation relevant and to avoid time wasting, it is important to avoid the verbal structures resulting from thought processes known as *communication roadblocks*.

Road blocks

Good conversation should consist of uninterrupted statements, expositions and questions all backed by good thought processes. However, our emotions often take control and some of the following communication blocks occur.

Judgmental blocks

- Criticizing   necessary at times, do not over use

- Name calling   avoid completely

- Diagnosing   use sparingly

- Praising evaluatively   often sensed as being false and a
                                 mechanism for manipulation.

Problem solving blocks

- Ordering   always causes resentment - avoid

- Threatening

  trying to control the actions of the other person by offering dire consequences - avoid entirely

- Moralizing

  telling the other person what he should do - usually causes anger

- Excessive

  usually leads to weariness, frustration inappropriate and often an abrupt ending questioning

- Advising

  if I were you I would do this ...etc. Try to advise only when asked for

- Directing

  talking about something which is of more interest to you than the other always causes irritation

- Logical argument

  an appeal to facts and logic will often have the effect of making the other person feel stupid, can be used but only with great care

- Reassurance

  sometimes helps, but usually sounds weak, never as good as on known positives.

Fortunately in the flying training environment, a high degree of motivation usually exists. If the trainer is not clumsy in his handling of the student, good quality dialogue can usually be achieved.

Observing body language

This is a highly developed skill area, but the trainer should endeavour to be aware of the general signs of stress, over-confidence and under-confidence. A student under stress will develop tunnel vision in his concentration and may miss something important. Watch for rapid breathing, white knuckles, clenched fists, body stiffening, fidgeting, missed calls.

A careful reading of all the signs can greatly assist your focus on problem areas. Looking at the face and eye contact will usually reveal a great deal about how a person is feeling. Listening to the voice is also a good guide. Research has shown that when someone gives a spoken message, the listener's understanding and judgement of that message comes from:

- 7% words - words are only labels, listeners put on their own interpretation on spoken words.
- 38% paralinguistic - the *way* in which something is said, by accent, tone or inflection.
- 55% facial expression - what the speaker looks like while delivering the message.

Postures and gestures - hands

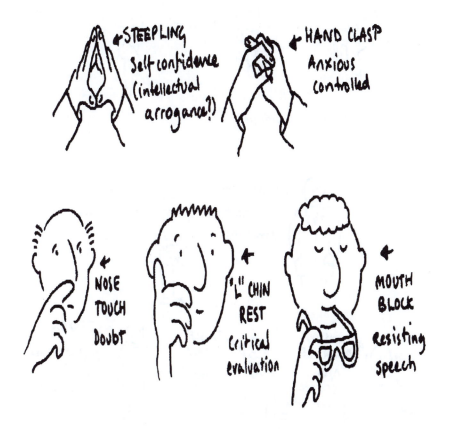

Postures and gestures - seated

Postures and gestures - standing

Voice signals

Sound of voice	*Feeling/meaning*
Monotone, voice	boredom
Slow speed, low pitch	depression
High voice, emphatic pitch	enthusiasm
Ascending tone	astonishment
Abrupt speech	defensiveness
Tense speech, loud to tone	anger
High pitch, drawn out	disbelief
High pitch, muffled, guttural	fear

Remember that although we can all see and hear another person's behaviour, the thoughts and feelings can only be guessed at.

The following chapter explores further body language behaviour and personality characteristics, offering some tips and advice in recognizing and dealing with difficult trainees.

The Captain's crew briefing, prior to departure is an essential team building exercise and sets the tone for the flight.

Effective use of audio visual equipment is of major value.

5 Dealing with difficult trainees

Very occasionally, especially in a group or classroom situation, the trainer may come across individuals who display characteristics that can disrupt proceedings. Here is what to do:

The heckler

Probably insecure
Gets satisfaction from needing attention
Aggressive/argumentative.

What to do:

Never get upset. Find merit, express agreement, move on, wait for a mis-statement fact and then throw it out to the group for correction

The talker/ know-all

Eager beaver chatterbox
Show off
Well informed and keen to show it.

What to do:

Wait until he takes a breath, thank, re-focus, move on. Slow him down with a tough question. Jump in and ask for group comment

The griper

Feels hard done by
Probably has a *pet grouse*
Will use you as a *scapegoat.*

What to do:

Get him to be specific. Show that the purpose of your presentation is to be positive and constructive. Use peer pressure.

The whisperers

Do not understand what is going on
Sharing anecdotes
Bored, mischievous.

What to do:

Stop talking, wait for them to look up and *non-verbally* seek their permission to continue. Sweep your eyes over the audience staying two or three seconds on each person, this gives the impression you are talking to each one personally.

    The silent one   Timid, shy, insecure
                                Bored, indifferent

What to do:
Timid - Ask easy question, boost his ego, bolster confidence
Bored - Ask tough question, use as helper in presentation.

Alternatives

If none of these methods are applicable the trainer may consider using *psychological judo*. In physical judo, you use the energy of your opponent to cause his downfall by changing your push into pull. In psychological judo, you ask the difficult trainee to be even more difficult. This gives them even more of the spotlight and attention than they wanted and they use all their energy to pull back to avoid ridicule.

It should be appreciated however, that the human characteristics listed above are generalizations, and that *some* individuals displaying these traits may not fall into the category of difficult trainees. *The silent one,* for example, may be an introvert, who learns by listening and reflecting on what has been said. *The whisperer,* may well not have understood. *The know-all,* may be able to contribute to the lesson, by giving him some space to contribute to the lesson.

Body language platform skills

Tips for classroom/group tuition

- Do not keep your eyes on your notes
- Never read anything - except extracts/quotations
- Exaggerate body movement and verbal emphasis
- Pause often - silence is much longer for you than for the audience
- Use humour
- Be enthusiastic
- Keep it simple
- Use eye contact
- Remember facial expressions account for 55% of understanding.

...body language!

Assertion

The trainer should approach his communication with the trainee using his own personality and natural style as far as possible. This will help to achieve spontaneity and assertion. The pitfalls to avoid are excessive submissiveness and aggression. Both these characteristics produce very negative results. The trainer should also avoid trying to be *Mr Nice Guy* or *Mr Perfect*.

When correctly applied the previously discussed skills of attending, listening and questioning can all be assertive. Assertive behaviour should not be misinterpreted as being over strong, forceful, or aggressive but rather a situation where you as the instructor feel good about yourself. You have a feeling of self worth which shows. This in turn generates confidence, both in yourself and the mind of the trainee. You will have positive control and your trainee will feel comfortable and secure with you. Knowledge of your subject and preparation for the role are key factors in being assertive.

'I' statements

When problems and concerns over training have been discussed, you will probably need to make some assertive statements. Again, knowledge of your subject is essential. There are five commonly used statements usually beginning with '*I*'.

- Expressing appreciation
  I appreciate the amount of effort you have given to this...
  I'm glad you saw fit to...

- Saying what you think without arguing
  *I think that we should approach the problem this way... I think
  that you should consider...*

- Setting limits
  *I am willing to go right through this again... Shall we fix a time for
  this later today?*

...assertive behaviour can be misinterpreted!

- Expressing frustration or anger
  *I feel that you should review this material thoroughly as you said
  you would... I have a problem with your approach to...*

- Confronting others when you have a problem
  *When you do X, I feel Y, because the effort on me is Z...
  I feel that you are not going about this task...*

In discussion, remain centred on the problem and do not be side-tracked or
enter into unproductive power struggles. If you are being attacked, remain
problem centred.

Guidelines

- Avoid defending yourself

- Avoid accepting blame as a means of terminating the discussion

- Avoid apologizing and offering excuses

- Avoid placating him as a means of terminating the discussion

- Remain calm under fire, do not try to change his mind by logical argument

...feelings can only be guessed at!

Conflict resolution

Inevitably, arguments and conflicts develop when two people have stray views on a subject. Heated argument releases individual *steam and tension,* in some cases, but is very weak communication. The problems are usually:

- A lack of listening
- An attempt at winning
- Lack of understanding

- Rigidity of mind, which prevents you from considering alternative solutions.

Leading to the possible outcomes of:

- You win, he loses
- You lose, he wins
- You lose, he loses.

A quick reference to conflict solution should contain the following three elements:

- Active listening

  Make sure that you do listen to his point of view. Make use of all your attending and listening skills, concentrate, sense his feelings, show that you are prepared to understand.

- Identify your position

  State your thoughts and feelings simply and positively without excessive repetition, or approach the same argument from another angle, which may reinforce.

- Explore alternative solutions - brainstorming

  List all possible solutions to the problem before evaluating their usefulness, do not be side-tracked. Then evaluate each proposal until an area of common agreement is reached. This method is not satisfactory in all cases but will usually allow a good airing of views and a maintenance of the basic relationship. It may be that at the end of the discussion a strongly held viewpoint is unchanged.

  If, after probing, alerting and challenging, you perceive there remains a problem with conflict resolution, then you may, as a last resort, have to exert your authority as a trainer and issue a warning.

Problem ownership

- Your trainee comes to you with a problem

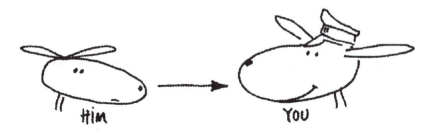

This is a counselling situation where you will need to use your attending and listening skills.

- You approach trainee with problem

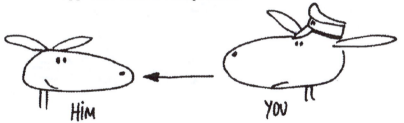

This is where an assertive approach should be used. Counselling here is inappropriate and ineffective.

- Both you and trainee have a problem

Use your problem solving, conflict resolution skills.

Giving assertive criticism

This skill will be necessary from time to time in your training programme.
It is a source of concern to many trainers but the concepts are simple:

- Always focus on the behaviour and not the person.
- Be objective, rather than judgmental.
- Do not attack and become aggressive although there may be strong
  feelings involved.

...assertive criticism!

Assertive criticism

- Describe
  What is the person doing that is causing a problem?
  Be clear and specific - do not use vague or unsubstantiated
  examples

- Express
  Why is it a problem? To him, to you, to the *system*

- Specify
  What behaviour change is wanted? How do we go about this
  change?

- Consequences
  Let the person know the positive consequences of changing,
  and the negative consequences of not changing.

Summary of assertive techniques

- Use active listening
- State your expectations clearly
- Have the ability to say no to a bad habit, practice a technique, do not be *wishy-washy*
- Give positive recognition where due
- Make good use of 'I' statements
- Always remain problem centred with positive focus on the solutions.

The next chapter focuses on how people learn.

Psychological skills training in leadership, assertiveness and human behaviour should form part of a pilot's induction.

Routine line training and checking gives the opportunity to provide feedback and assess the crew operation in the real world. Line training however, is pointless if selection on hiring has failed to show that pilots have aptitude.

6 The process of learning

Learning is a process of acquiring new facts, skills and attitudes, causing a change of behaviour as a result of experience. The ability to learn is one of the human species most outstanding characteristics. Learning occurs continuously throughout a person's lifetime. Because of a learning experience, an individual's way of perceiving, thinking, feeling and doing may change.

Principles

Learning is a very complex process. Psychologists have listed various principles that are reasonably constant in all learning situations. In the study of learning, these principles do not apply dogmatically, but flexibly, both in order and statement of content. The generally agreed principles of learning are:

- Readiness
- Exercise
- Effect
- Primacy
- Intensity
- Recency
- Special interest.

A study of these items will reveal the ways in which people learn, generally referred to as the *principles of learning*, as well as the role of recall, application and memory.

Readiness

This principle, states that individuals learn best when they are ready to learn and they do not learn much if they see no reason to learn. Readiness implies a degree of single-mindedness and eagerness. When trainees are ready to learn they meet you at least half way and this simplifies your task. Under certain circumstances you can do little, if anything, to inspire in trainees a readiness to learn. If outside responsibilities, interests or worries weigh too heavily on their minds, if their schedules of living are overcrowded, or if their personal problems seem insoluble, they may have little interest in learning. In these situations, you can only help, in perhaps identifying the factors causing lack of readiness.

Exercise

This principle, states that those things most often repeated are best remembered. It is the basis of practice and drill. The human memory is not infallible. The mind can rarely retain, evaluate and apply new concepts or skills after a single exposure. Every time practice occurs, learning continues. You must provide opportunities for trainees to practise or repeat and must see that this process is directed towards a goal. Practising the golf swing in the home garden is rarely as effective as practising the swing in an actual match on a golf course, with a real target green to aim for. A similar principle exists in flying and the trainee should see practice as a process of continual improvement. Positive reinforcement is also necessary. When practising landings, for example, you should not only mention the things which went wrong with the last landing, but the things which must be done correctly to achieve a good landing next time.

Effect

This is based on the emotional reaction of the learner. It states that learning is strengthened when accompanied by a pleasant and satisfying feeling, and that learning is weakened when associated with an unpleasant feeling. Again, you have a great deal to do with the principle of effect. You should be careful not to introduce difficult tasks too early in the programme, leading to an experience, which produces feelings of defeat, frustration, anger, confusion or futility.

Positive attitudes are important. Instead of emphasizing or stressing the difficulty of a task, you should point out that although difficult initially, correctly applied procedures and practice will soon make it easy. It should also be stressed that the problem area is well within the trainee's capabilities to understand or perform. It has been found that introducing pilots to automation is one area where the principle of effect applies. Correct sequence of instruction and a step by step approach will help avoid frustration.

Summarizing

- If a subject is easy, instruction should come early, leaving plenty of subsequent periods for testing and practice

- If a subject is difficult, it should be introduced gradually and spread more evenly over the whole training programme with regular practice sessions

- Both instructor and student should expect to make slow progress at first on a difficult subject, and expect to find that performing an easy task may even become boring

- Unless a trainee is adequately warned, his personal expectations may be too high, and when a steady rate of progress is not achieved, he may become disgruntled and disappointed.

Primacy

This states that first impressions are always important, this is particularly so in a learning situation. For you, this means that what is taught must be right the first time. For your trainee it means that the learning must be right. *Unteaching* is more difficult than teaching. Try to start well, on a positive note and lay a good foundation for all that is to follow.

Intensity

A vivid, dramatic or exciting experience learning something teaches more than a routine or boring experience. This implies that a trainee will learn more from the real thing than a substitute. Obviously some learning situations impose limitations on the amount of realism obtainable. This should present a challenge to your imagination. It is in this area that a great deal of effort has been spent by airlines all over the world on training aids to heighten the intensity and effect of learning. The following is a list of aids, which are available and should be considered for use by both you and your trainee.

Simple aids	*Teaching machines*	*Sophisticated aids*
sketch pad	audio visual trainer	cockpit systems
white board	instrument flying trainer	flight simulator
charts	CBT	supernumerary flying
slides	navigation simulator	line flying
movie film	training simulator	
diagrams	cockpit mock up	
posters	tape recorder	
photographs	video recorder	
model aeroplanes		

Recency

This states that things most recently learned are best remembered. The importance of this, as far as instruction is concerned, is that we need to repeat, restate, re-emphasize and summarize. It is often helpful after a break in flying training to recapitulate key learning points, before commencing the next flying session.

Special interest

This principle of learning does not always apply but sometimes accounts for a sudden surge in learning progress following a slow start or a period of stagnation. It can often be responsible for creating a positive motivation for the whole learning programme. This theory applies particularly to the trainee who is fascinated by a strong narrow interest that is capable of being enlarged. For example, some pilots love *gadgets*, and when they meet computers and automation in their training, a strong motivation is released, which usually has the effect of increasing enjoyment of the whole programme as the individual integrates his special interest learning.

...first impressions are important!

How do we learn

Psychologists studying learning behaviour have listed principles, which they suggest be present for any learning situation to progress to a successful conclusion.

These can be stated as:

- Perception and stimulus
- Motivation
- Repetition
- Application
- Participation
- Logic
- Transference
- Use
- Interest
- Expectation.

However, learning may be accomplished at any of several levels, some of which do not appear to use all of the principles listed above. Short cut or quick method learning can be accomplished in a variety of ways, thanks to the wonders of the human brain *(refer to Chapter 7, The brain - memory)*. Memory alone is a very complex subject. Factors such as short-term recall, long-term recall, random access memory, the subconscious and even hypnosis, are all an important part of the learning process.

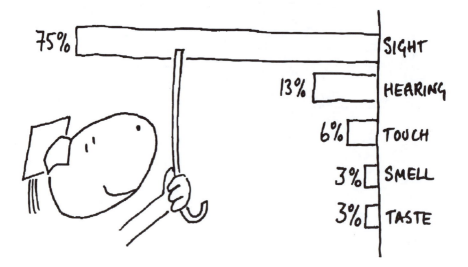

...most knowledge acquired through sight!

Perception and stimulus

Learning is stimulated through the following senses:

Sight	75%
Hearing	13%
Touch and muscular sense	6%
Smell	3%
Taste	3%

Sight is the master sense. TV, video and visual advertising, evidences this fact. In flying, the kinaesthesia sense is important as it is involved with physical co-ordination and three dimensional judgement and timing. It is closely related to touch. Good quality learning requires the simultaneous use of as many senses as possible.

Helping people to learn requires in part, finding ways to aid them in developing better perceptions. It takes time and opportunity to perceive. This time varies greatly with individuals, and the result of perception is directly related to the basic needs of the individual. Perception of the same object, the famous *inkblot* test and the psychiatrist's question, *tell me what this reminds you of*, will give different results when shown to a group of individuals and, of course, we are all subject to sensory illusions, a fact heavily relied upon in the use of flight simulators. The art of flight simulation almost requires a pilot to use his imagination and *go along with the illusion* of the senses to mentally *think* he is flying. The more complete this illusion, the better the simulator is as a learning tool. This submission to the illusion of sensory perception has been highly successful, as flight simulators are now considered better teaching machines than the real aircraft they represent.

It is important for you to remember that differing levels of perception of a seemingly simple fact do not necessarily indicate differences of intelligence, but rather the different needs within the individual. The successful instructor matches the subject to the need in the most effective way.

Motivation

Motivation is a very strong force and may be either positive or negative, tangible or intangible. Negative motivations are those that engender fear or threats, and are not usually a factor in flying training. Positive motivation is more obvious and is the promise of achievement or rewards, giving satisfaction, financial gain, improvement of self-image and public recognition.

The desire for personal gain, either acquisition of materials, or position, is a basic motivation for all human endeavours. An individual may be motivated to dig a ditch, or to design a supersonic aircraft, solely by the desire for financial gain. Positive motivation is essential to true learning.

...mentally think he is flying!

Fortunately, in the aircraft industry, the sense of wonder and fascination about aircraft and flying, is very strong motivation for all levels of learning, among personnel involved.

Most aircrew are motivated to do the job well. Less than satisfactory performance is easily detected by their peer group, by passengers and superiors. As in many endeavours, true satisfaction comes from the motivation inherent in satisfying the old cliché - *something worth doing is worth doing well.*

Repetition

Repetition can reinforce short-term learning but too much can have a negative effect - remember how quickly irritation builds up when the needle on a record track *sticks*. Repetition coupled with practice is a foundation of learning and skill development. No amount of telling will teach a man to swim, or to fly an aeroplane, he must do and then practice. An instructor needs to plan and space his repetition - this is where the *principle of application* can be applied.

- Tell them what you are going to tell them
- Tell them
- Tell them what you have told them
- Ask them to tell you what you have told them

Rote or drill methods can also be used: for example, in learning numeric facts or limitations.

...whenever possible use humour!

TV commercials are examples of a method of repetition, which although it can be irritating, does get results. Individual advertisements have even been known to develop a cult following. Repetition is more effective when the following guidelines are observed:

- The primary stimulus is varied on each repeat - tell him, show him let him tell you, then show you.

- Encourage a better performance on each repeat.

- Do not show irritation, annoyance, or sarcasm.

- Whenever possible use humour, *not directed at your trainee*.

- Remember that a task that has been repeated carelessly or incorrectly, forms a habit that is hard to break.

- Praise and reward successful repetitions. Show your pleasure in the achievement of a correct result. Smile.

- Expect the possibility of a plateau in the learning curve, i.e. a period of no apparent progress, often followed by a small gain and then another plateau followed by a large gain.

...practice makes perfect!

Remember that practice in itself will not make perfect, but practice together with a feedback of information on the results of the trainee's efforts, particularly when successful, will certainly tend to make perfect. Swinging a golf club, hitting the ball and watching its flight provides feedback. When the instruction is correct, the practice is correct, the feedback is positive, the learner receives a reward, this completes a loop towards rapid progress.

Spaced learning

More may be learned in three spaced learning periods of 30 minutes each than in one period of 90 minutes. The dosage in any subject is related to muscle fatigue and nervous fatigue (neuroma). Some form of learning or adjustment takes place between training periods (during the rest period). The chance to *chew the fat*, and hold discussions with contemporaries, or peer group trainees, after a training period, is very important consolidation.

Most people have experienced the effect of clearer perception of a problem following a good night's sleep. What seemed difficult in yesterday's practice suddenly becomes obvious. The message here is, do not labour the point and indulge in excessive repetition.

Application

Immediate application improves learning. Make use of motivation and short term enthusiasm. Allow *hands on* training whenever possible in any learning situation. Any lesson presentation should normally follow the sequence of *preparation - application - review - evaluation.*

The timing and presentation of a newly developed skill is of great importance in successful learning. The best training removes all surplus content. The trainee learns what he *needs* to know to do the job. Application must be meaningful. What we learn rationally, we remember, because we have understood.

A person with no knowledge of the principles of flight can be shown how to turn an aircraft - *just turn the wheel and push on the rudder pedal!* His short term application will not stand up to the varying conditions of aircraft flight, compared to a person who has been shown the same actions, but has learned and understood why these actions will produce the desired result. Something that has no meaning is isolated, it tends to drift away, whereas that which has meaning is tied up with other things.

Associated with this is the fact that we are apt to attend most closely to those things that have a bearing on what we already know. In application of learning there is a need for you to develop your trainee's subject from the known to the unknown, to tie his new facts into the knowledge he already possesses.

Participation

Active participation is the difference between learning about something and learning to do something. There are many *experts* on, for example, football that may know more about the game than the players, because they watch the game avidly. However, they do not have the skills to play the game. Each field of endeavour, including flying, has its *armchair experts* who learn by quite legitimate methods, but generally do not actively participate.

This subject is very much involved with good teaching technique, as we can have the situation where even though a trainee is in a learning group, he is not included by the instructor, and so learns from peripheral activity only. Active participation is trying, doing, repeating, discussing, working problems, exercises and asking questions.

...meaningful application!

Logic

The principles of logic state that each point in a learning situation must make sense by itself and learning points must come in logical sequence.

There are four types of logic:

- Cause and effect
  If you forget to lower the undercarriage before landing ...!
- Generalization
  Generalizations are often weak learning points, as sweeping statements of fact tend to be made from an insufficient data base
- Relevance
  Going from the general to the particular
- Analogy or comparison
  Things alike in some respects will be alike in other respects. This is very heavily used in learning.

This is the *for example* technique. Trainees often learn more from the examples, in terms of real understanding, than from the statement of the rule.

Analogy and comparison offer a vast field for you, allowing insertion of tall stories, humour, mild reprisal.

...active participation!

Sequential events and time frames

As flying is a sequential activity, with a high degree of repetition, a great deal of learning is achieved through the method of what *comes next,* or *it is time to do this.* The very nature of the sequence, start, taxi, take-off, climb, for example, encourages *trigger points* initially to jog the memory, and subsequently to establish learning habit patterns. These *trigger points* are very strong and can be used to good effect by trainers. Their use, however, should not be overdone otherwise, they tend to become performance *crutches.*

Where possible, they should be related to performance of a task which involves a strong safety point, for example, *I do this only when I'm clear for take-off* or *I do this only when I'm clear for final,* or *I do this to remind me of an altitude restriction,* or *I do this to remind me of fuel cross-feed* and so on.

Similar triggers are activated not by position but by time. These are equally powerful and effective. Sequential events and time frames even affect hunger and thirst responses in crews. Some pilots are so conditioned to having a cup of tea or coffee, following the after take off drills, that they may appear to suffer withdrawal symptoms if for some reason the drink is not available, even though there is no logical reason for a drink at that stage of the flight.

...we learn but we pass it on!

Transference

This type of learning depends upon comparisons and the transfer of similar or common ideas. It can often be used to develop a difficult concept from an easy concept and to lead on from previous knowledge.

> *A jet engine is like air escaping from a balloon .. in some respects!*
> *Using an ILS is like using a VOR .. in some respects!*
> *A Boeing 767 is like a Boeing 737 .. in some respects!*

Like motivation, transfer can be positive or negative. If learning A helps to learn B the transfer is positive. If learning A hinders the learning of B, negative transfer occurs. Transfers sometimes lead to the so-called *mental block* particularly when negative transfer is involved.

A pilot may make consistently bad landings on one particular runway, for no apparent reason. This quite quickly builds up into a negative transfer. It helps here to use diagrams, pictures, whiteboard sketches. These may uncover the problem area. One successful result will usually remove all negative thoughts. The same problems occur in other highly developed skill areas, such as tennis or golf, where one good shot can lift a whole game.

All pilots will have experienced a run of bad landings and get a positive mental lift when, at last, the good one comes, particularly if they feel that they have identified a problem area. In developing thought processes, most thinking is sequential, using transference of the *building block* type. Sometimes thinking processes are both vertical and lateral. Lateral thinking is often innovative and creative thought.

Usefulness

All learning should be useful to the trainee. He should also be able to see the use of the material. This leads to an economy of effort, avoidance of duplication and waste. In short the *need to know* concept as opposed to the *nice to know* concept.

Interest

In effective learning, the senses must be continually stimulated. Motivation should not occur just at the start of a project, but all the way through. When explaining something, you should use your best explanation first, for example, the way in which you find it easiest to understand. This may or may not interest your trainee sufficiently, and a new explanation may have to be used and tested for understanding. In order to maintain interest all the previously listed learning methods are used but saturation must be avoided.

A comedian has failed when the audience loses interest. The subject matter itself should stimulate sufficient interest, but sometimes you have to supplement this by *showmanship* in the learning situation. Maintenance of interest is a very difficult area in this role and you should be careful not to set impossibly high achievement targets.

...motivation!

Expectation

In any learning situation involving an instructor, a level of satisfactory performance is reached more easily, if the trainee is aware of what that level of satisfactory performance is. Expectations should be clearly set, so that there are no open ends and the trainee can honestly say, yes, *I can do it and I can do it well.* In a line orientated training session, for example, this can be achieved by angled questions that result in the trainee making his own assessment of his performance, at the de-briefing.

Characteristics of learning

Each trainee sees a learning situation from a different viewpoint because learning is an individual process and not a group process. In a demonstration of a particular sequence in an aircraft, the trainee may not learn what you had intended as the prime learning goal of the exercise. He may have been affected by other more vivid or intense feelings, he may be distracted or even through apprehension he may not have learned at all.

A common example of this feature of learning occurs in light aircraft training when the instructor demonstrates a fully developed spin, intending that the trainee should learn the recovery technique of *opposite rudder and forward stick*. The trainee often finds the experience so vivid and unusual, that his learning experience is all to do with an apprehensive feeling, visual impressions, extreme attitudes, sounds and heightened awareness. This intensity can initially obliterate previously memorized drills, which can only be regained by careful handling on the part of the instructor and multiple repetition of the exercise.

In the airline flying situation, shock and surprise can erase the memory of a Phase 1 emergency drill, but after a period of recovery, a correctly learned procedure can still be recalled provided that the characteristics of vividness, intensity, and repetition have been emphasized in the learning process. In this area of learning, flight simulators give very valuable experience.

Similarly, distraction can be damaging in the learning process. The trainer may give a perfect demonstration of manipulation and *patter* on an ILS approach, intending that the trainee should learn the appropriate configuration changes, cross-checks and calls as well as appreciating the appearance of the runway at the minimum. If however, the trainee's seating position is incorrect, or if he has not been briefed on auto throttle movement, or aural voice warnings, he may well be distracted and miss the finer points of the exercise and may not even see the runway at all. You need to use many of the principles of learning in combination with sufficient preparation and attention to detail to ensure a worthwhile result.

Learning through experience

The instructor cannot do it for the trainee. A person's knowledge is the result of experience and no two people have had identical experiences. Even when observing the same event, two people react differently. They learn different things from the experience according to the manner in which the situation affects their individual needs. Listening to two people describe the events of, for example, the same football match is often very enlightening from the point of view of the way the experience affected them as individuals.

- Previous experience conditions a person to respond to some things and ignore others

- All learning is by experience, but takes place in different forms and in varying degrees of richness and depth. For instance, some experiences involve the whole person, others only the ears and memory

- Learning lists of words is a rote learning experience which we all need to use from time to time, but the memory drill alone is not very meaningful

Consider a pilot who has made a list of the systems affected when the fire shut-off handle is pulled in his aircraft, and attempts to commit this list to memory. How much more vivid is the learning experience when he actually pulls the handle either in the aircraft, or simulator, and is able to see, feel and hear what happens. Similarly, no amount of reading about the use of radar and turbulence penetration technique in thunderstorms equates with the real experience. Most pilots get quite a surprise initially, when they watch a propeller being feathered after they activate the feather button, even though they know what is going to happen, and have read the appropriate sections of the aircraft operations manual.

- It is a more vivid experience if an experience challenges the learner. Seeing, doing and touching is a better learning experience, requiring involvement with feelings, thoughts, and memories of past experiences and physical activity. It is more effective obviously, than an experience in which all the learner has to do is commit something to memory. Modern teaching machines usually incorporate a high level of different learning techniques.

- Good instructors always seek ways of providing learning experiences that are meaningful, varied and appropriate. It seems clear enough that the learning of a physical skill requires actual experience in performing that skill, but many more facets of learning come into play when we are learning how to do something. This is referred to as multiple channel learning.

Learning through multiple channels

Learning skills involves using more than memory and muscles. Learning can be broken down into a number of classifications that usually operate simultaneously, to achieve not only the primary learning task, but incidental learning as well. For example, when conducting an instrument approach in a learning situation, the following *channels* will be in operation:

- verbal learning
- sensory perception
- conceptual learning
- motor skills
- problem solving
- transference.

...memories of past experiences!

At the same time, incidental learning is taking place, including:

- Development of attitudes about aviation, procedures, discipline, habits and mannerisms

- Management skills

- Self-reliance.

It is clear that as a trainer you are a very important person to your trainee, during the duration of the learning period, and that your influence is stronger than sometimes suspected. The process of learning is active and continues subconsciously long after flying duty hours are over.

The nature of skill

Skill involves:

- *Perception* - The gathering of information
- *Processing* - The making of a series of decisions
- *Action* - Putting the successive decisions into effect.

...a very important person!

Each of these steps requires both time and opportunity to carry out. All the previously discussed principles of learning apply. The rate of learning a skill varies considerably from one individual to another. The reasons for these differences are very complex and so generalizations have to be made for the purposes of our training requirements:

- There will be differences in performance, in skill, depending on the needs and aptitudes of the individual. For example, in learning to ski on snow, some people take only minutes, some days, and some never achieve basic skill on skis. The average time taken to achieve basic self-sufficient skill levels is about one week of applied practise. On the other hand, of course, the person who never succeeds on skis may learn to play the piano without any difficulty. Any skill learning depends on the needs of the individual. When the need and the aptitude match, the skill will be acquired. The level of performance will vary again with the level of motivation and practice. Some people will feel moved to practice a skill for a few days whilst others will practice a skill for a lifetime

- Learning a difficult or new skill takes a very high percentage of a person's concentration, and initial rate of progress is usually slow. This fact requires patience on your part. It also means that your trainee sometimes has very little reserve capacity to handle any other problems at the same time.

- In learning a skill, short periods of applied practice with rest periods in between, will result in the most rapid rate of progress. As the level of skill improves, more reserve capacity will be released.

In airline flying, we are dealing with pilots who already have basic flying skills at the ab-initio level, and highly developed flying skills at the advanced level. This means that our skill learning mainly comprises an adaptation to new techniques in a related task. However, as a general rule the more advanced the aircraft, the more a pilot needs to draw on his skill and experience, and the probability is that he will still need to acquire new skills in order to be successful.

For example, a pilot converting from propeller driven aircraft to jets, needs to acquire an understanding of several different concepts and skills, for example, swept wings, engine spool up times. A pilot converting from the Boeing 727 to the 737, for example, needs to adjust to a different thrust-line. A pilot who is converting to a Boeing 747, needs to be aware of the differences in cockpit height at the flare. Other factors have a strong influence on skill learning as well, and include:

- Cockpit environment
- New aircraft conversion and initial command together
- New aircraft, new route structure, or new role.

...circling approach procedure!

A pilot needs to constantly practice all his skills. Instrument flying in particular needs consistent application and practice. Sometimes, for various reasons, this practice is not always able to be applied at the desired level, so you should always be prepared to recognize a skill problem and be prepared to retrain an individual where required, from quite basic skill levels, if necessary.

A person who is performing a skill at a high level will generally be relaxed, confident, and able to use some of his reserve capacity to handle other tasks, for example, radio calls, checklists, and abnormalities, without detriment to the performance of his primary task. As skill level falls off, the primary task will take more and more of his capacity.

If this happens, the pilot will suffer a reduced rate of:

- Perception
- Decision making
- Trend recognition
- Primary action
- Corrective action.

Learning, or re-learning a skill requires the following elements to be present:

- Discussion and presentation of a pattern. A step-by-step approach to the skill
- Demonstration
- Repeat demonstration
- Practice
- Comment, praise, evaluation
- Further demonstration to correct errors
- Further practice
- Review, evaluate, rest
- Further practice after rest.

As skill develops: *anticipation* will grow
                     reaction time will speed up
                     recognition of trends will occur earlier
                     judgement will improve

accuracy will increase

feedback will be more meaningful

problem resolution will be more successful.

In evaluating a skill, you have to ensure that the required inputs are correct even though the person may appear to be achieving the required standard. This will ensure that errors are not being practised and reinforced and that advancement to a higher level of skill will be built on a sound foundation.

Behaviour change

At the start of this chapter learning was defined in terms of a change in behaviour. The primary purpose of any education and training program, therefore, is to produce learning, i.e., to achieve a permanent change in behaviour or performance.

Learning to fly, upgrading to a new aircraft type or migrating to a new platform involves changing one's knowledge, understanding and practical skills. Whenever we learn we either have to acquire something new or change something we already know, or both. In an age of rapid technological progress and legislative flux, the next change in equipment, operating environment and procedures, is just around the corner. Clearly, the need to re-learn, to be flexible and to change and adapt quickly is a constant challenge in today's world.

The problem with this, though, is that people dislike and resist change. We all prefer our comfortable and familiar surroundings, procedures and routines. Having invested all that time and effort to practice and learn something and become skilled, it is much easier to stick with what you know than to face having to change and start all over again.

But everyone has to change, sooner or later. You either *have* to change or you *choose* to change. On the one hand, change may be forced on you because your procedure or technique is sub-optimal, incorrect and possibly unsafe or because of some other pressing requirement to improve. On the other hand, you may choose to change not because you are doing anything wrong but because you want to upgrade, transition or change operating platforms. Whatever the reason and whether it is forced or unforced, change presents you with the task of learning new knowledge and skills and changing the old.

But people keep falling back to old ways!

Teachers and trainers as well as their students all try to get it right the first time but trainees often get it wrong. Inaccuracies, imperfections, misconceptions, shortcuts and errors inevitably creep in, despite quality training and highly motivated students. For whatever reason, whether it be inattention, fatigue, loss of concentration, work pressures, carelessness, distractions and so on, people end up not following standard operating procedures. Poor or unsafe working practices can develop which generally go on uncorrected and soon become established, routine ways of working and performing. A familiar and frustrating outcome of training programs is that people often fall back to old ways soon after completing their training. For example, when learning to fly or qualifying for a new aircraft type, the trainee may appear to be able to adopt the desired behaviour during the training session, but once in the air much of what they apparently were able to do during training seems to disappear. Trainees in this situation invariably revert to their 'own way' and forget their training. *(see Chapter 7, The brain - memory)*

Reversion to old ways is not likely to occur while the trainee is under close supervision or is concentrating hard on what to do but it is much more likely to happen during moments of high activity and heavy workload or when concentration lapses. Unfortunately, reversion to old ways is likely to strike when you can least afford to make a mistake and when the situation demands that you respond in the way you were trained; not as you used to do.

What's the answer?

Traditionally the answer has been practice and, when this fails to improve performance, re-training. Whilst this works to a degree, it is somewhat inefficient and not cost-effective. It does not tackle the root cause of the reversion problem and the slow rate of learning progress associated with interference from the learner's prior knowledge. How many times have we heard someone say when converting to a new type 'it's not learning the new – it's forgetting the old'?

Because our established performance routines are automatic and therefore beyond conscious awareness, it is extremely difficult to change old ways. It is generally accepted within the aviation community that it is easier to alter equipment, tools and the work environment than to try to change work habits.

...self-improver!

What do we do about non-compliance or non-achievement?

When training does not produce the desired changes in behaviour and performance and the quality of the training is not in question, we typically resort to re-training as a solution. Clearly, if the person did not learn anything from training the first time, then we need to repeat it.

This begs the question, 'How do we know that the person did not learn anything from his initial training?' The answer is that he must have forgotten what he was taught because he keeps falling back to his old way - what he used to do before he was trained. The old way may be 'doing the wrong thing' or 'doing nothing at all - failing to act'. Both are instances of non-compliance and both suggest that the initial training, for one reason or another, has failed.

Having decided that re-training is necessary, we put the person through the course again and wait and see if the desired behaviour changes occur, this time. If non-compliance and reversion to old ways resurfaces again, we may start to overtly blame the person and get into an authority relationship and try to coerce compliance. This produces avoidance behaviour and diminished performance and the poor or unsafe working practices simply continue. The person may become progressively disengaged with the job and may start to blame himself and confidence suffers. All these are undesirable outcomes of the initial failure to learn and the associated non-compliance. We need to de-personalise all this.

Why don't people change? Why do they keep falling back to old ways?

People may be new to a training situation but they are definitely not a 'blank slate' for the trainer to write to. Even beginners have some knowledge, some ideas and some self-taught skills. Trained pilots have much knowledge and many skills.

Some or all of this knowledge and pre-existing skills that people bring to the training situation may be incomplete, misguided, misinformed, misconceived or sometimes completely incorrect. These 'wrong' ideas and ways of performing, having gone uncorrected for some time, have been practiced (repeated) and have now become ingrained, routine, habitual, automatic, reflex-like responses to particular situations and instructions.

We know that it is possible to learn the wrong thing just as easily as you can learn the right thing. If you practice the right understanding and the right procedure, *that* is what you will learn and be able to repeat next time. Similarly, if what you practice is the wrong understanding and the wrong procedure, then *that* is what you will learn. That is how misconceptions and poor or unsafe working practices and procedures develop into learned, automatic behaviours. You then have a 'learned error', also known as a 'habit error'.

Learned errors, like all habits, die hard. Unfortunately, conventional training and re-training programs take little or no account of a person's pre-existing learning, whether it is correct or incorrect. We either assume the trainee knows nothing; or that he knows enough to benefit from the next stage of learning. These assumptions are seldom verified; we simply assume that because the person either possesses or does not possess a particular training qualification that they have or do not have the pre-requisite knowledge. However, we ignore the learner's prior knowledge at our peril.

Prior knowledge matters

It has been suggested that the single most important factor influencing learning is what the learner already knows. Ascertain this, said the psychologist David Ausubel, and teach accordingly.

We said earlier that when someone makes an error, i.e., takes the wrong action or fails to act (non-compliance), it is usually taken as a sign that the person has *not* learned something. This apparently logical assumption is based on initial observation but, in fact, it may be quite wrong. Let us take a look at what actually happens in many non-compliance situations, using a useful diagnostic process called error analysis.

Error analysis: diagnosing non-compliance and performance difficulties

It is generally accepted that many errors during learning result from careless mistakes, inattention or inexperience and these kinds of errors usually disappear as the person gains mastery. Let's call these *Type 1* errors. Type I errors are random, inconsistent and show no obvious pattern.

Unlike the random errors made by beginners, errors made by more experienced performers do show a pattern - they are consistent, repetitive and clearly not due to carelessness and inexperience. These 'expert errors' are also persistent in that they resist most eradication attempts. They are examples of what are called 'habit errors', 'learned errors' or 'interference errors'. Let's call these *Type 2* errors.

...rehearse, rehearse, rehearse!

While the inconsistent Type I errors indicate the absence of learning, habit errors are a sign that the person *has* learned something - he has learned how to do it 'wrong'.

Error analysis can help identify which kind of error we are dealing with by looking for a pattern or consistency in performance. Research tells us that human errors often reveal a surprising degree of consistency. This goes against the widely held mistaken belief that most errors are random, chance events or simple transient mistakes.

The consistency of most errors has two critical implications for aviation training and also for all other kinds of education and training, namely that:

- learned errors or habit errors are much more widespread than previously believed; and
- Type 1 and Type 2 errors require quite different teaching approaches.

In actual fact, though, most trainers treat these errors as if they were the same and therein lies the problem.

Re-training or re-teaching is not always the best solution

Type 1 errors

- Indicate that the desired learning did *not* occur. Perhaps the person was inattentive, careless, distracted or unmotivated or the teaching method was inappropriate or the learning was unsuccessful for some other reason. The correct remedy is to repeat the teaching because it did not 'take' the first time. Re-teaching or re-training using appropriate compensations and adjustments is the right solution.

Type 2 errors

- However, indicate that learning *did* take place but instead of learning the 'right' way, the person unfortunately learned a different and 'wrong' way. What he learned may be wrong and indisputably wrong but since it is learned it is automatic, habitual and persistent. Re-training does not work well with Type 2 errors so it should not be used.

Selecting the training approach with the best chance of success requires an initial error analysis. However, this is not generally done, for several unfortunate reasons.

Conventional training methods do point out errors if they are picked up and after that the training process spends no more time on the error and instead emphasizes the correct answer. Because errors are believed to indicate the absence of learning, i.e. failed learning and by association possibly failed teaching, trainers prefer not to dwell on errors any more than necessary. This predisposition is also motivated by the mistaken belief that paying too much attention to an habitual error will make it even more difficult to eradicate.

What the trainer is basically saying to the trainee during error correction or correction of non-compliance follows the conventional best-practice approach to training that is familiar to us all. It goes something like this:

> *'That's wrong; don't do it that way because ... ; I don't want to see you doing it wrong.'*
> *'Do it this (correct) way - watch me ... listen to my explanation...'*
> *'Now, copy my action/repeat what I said.'*
> *'That's good but try to pay more attention to ...'* (constructive feedback)
> *'O.K. Now go and practice it and I'll see you again next week.'*

Prescribing practice of the 'right' way can be effective over an extended period of intensive skill correction but learning gains are typically slow. Most importantly, transfer of learning to performance settings outside the training setting is usually poor because whenever the trainer is not directly attending to the trainee, the cues to correct performance are removed and so the trainee typically reverts to his old, wrong way.

Type 2 errors indicate that we are dealing with an *unlearning* situation rather than a straightforward re-teaching situation. Incorrect prior knowledge stands in the way of learning progress and it cannot simply be ignored or glossed over; it must first be corrected, i.e. *unlearned*. Unfortunately, *unlearning* old ways is notoriously difficult using conventional training methods because 'old habits die hard.'

Unlearning of misconceptions is also necessary

The problem of undetected errors being practiced and then developing into learned, habitual, ingrained errors applies not just to practical skills but also to attitudes, beliefs, ideas and conceptual understanding. Many of our misconceptions, misunderstandings and incorrect beliefs are ingrained and habitual and very resistant to change, even in the face of clear evidence to the contrary.

For example, FAA research on crew-automation interfaces shows that the first year on new aircraft has much trial and error learning, even for pilots transitioning from one automated aircraft to another. Many of these performance difficulties appear to be conceptually based in that pilots' understanding of the flight management system is often faulty. Even experienced and qualified pilots often have misconceptions or incorrect or incomplete understandings about how the flight management system works. Some mistakenly believe that it makes one decision at a time even though it is actually capable of making many decisions simultaneously. The pilot therefore often fails to effectively use the automated system because he underestimates its capabilities. These misconceptions, like other learned errors, arise despite initial training and are known to be hard to correct using conventional re-training methods.

Furthermore, re-training aimed at using the flight management system better, e.g., implementing a rational or discretionary use strategy, has been shown to be ineffective while these misconceptions persist in the minds of pilots.

...can't teach an old dog new tricks!

Learning something new often means first *unlearning* the old

While it is true that learned errors can present great difficulties for trainers and trainees, it is also true that correctly learned skills can also eventually become a barrier to further progress. Pilots undergoing upgrade or transition training often find they first have to *'unlearn'* their existing skills in order to acquire new skills. This *unlearning* process can sometimes be frustrating, time consuming and expensive.

Once again, it is a case of the pilot's prior correct knowledge becoming a learned error – 'yesterday in my old aircraft it was quite alright to do it that way, but today in this new aircraft the old way has become the wrong way and has to be *unlearned*, and unlearned quickly'.

Sooner or later every pilot comes to a point where it becomes necessary to change his or her skill base. If the new skill is consistent with the old skill, then the new simply builds on the old and change is rapid, uncomplicated and relatively stress free. But if the new skill is different from and conflicts with the old skill, as in the case of trying to correct a learned error or when a technique or procedural change is required, then it will be hard for the pilot to change quickly. In fact, it can take up to 2,000 repetitions of a new skill before it completely replaces a pre-existing old skill. Clearly, if you simply practice a new skill over the top of an old one that is already there, it will take a long time before you overcome the previously learned behaviour and finally start to improve to the stage where you are fully comfortable (automatic) and competent in the new skill or procedure.

Prior knowledge and skills can either enhance or interfere with subsequent learning. If pre-existing learning is correct and is consistent with new learning, then the old enhances the learning of the new. However, if prior knowledge is different from the new information being taught, either because it is updated, more efficient, safer, more cost-effective and so on, then the brain immediately detects this conflict and involuntarily activates a brain mechanism called *proactive inhibition*, also known as proactive habit interference.

Your old learning can interfere with new learning

Proactive inhibition (PI) is actually a knowledge protection mechanism. PI can be described as forward acting interference with learning such that 'old learning interfering with new learning'. PI is the underlying cause of most non-compliance, non-achievement and apparent failure to learn.

PI's real function is to automatically protect any attempt to change our prior knowledge. PI preserves all our prior knowledge against the onslaught of change demands. Without PI we would face having to relearn everything from one day to the next.

Unfortunately, because PI does not 'know' whether our current knowledge base is 'right' or 'wrong', it equally protects incorrect, as well as correct knowledge and skills. This is why old habits die hard. This is why we find change so difficult, uncomfortable and frustrating.

PI works by accelerating the forgetting of new knowledge and skills whenever these conflict with prior knowledge and skills. Telling a person he or she is wrong, however well-intentioned and received, only serves to activate PI which then inhibits retention of the new learning the teacher is trying to impart. Within a matter of minutes or hours the learner appears to forget what he has just been taught and reverts to his old incorrect knowledge and skills. This is known as accelerated forgetting. It is one of the main reasons why people return to their old ways soon after being taught something new, despite being well motivated to change. It also explains why pilots can pass tests and examinations, yet afterwards still not perform at the level of their qualification. PI is involuntary and we have no control over it. Everyone has PI but some have more of it than others hence some individuals adapt to change remarkably easily while others find it hard going. Simply being aware that you have proactive habit interference will *not* help you overcome it.

The most important thing to realise about PI is that conventional training and re-training methods actually increase proactive habit interference. This unintentional outcome is an unfortunate side-effect of the way we normally teach and learn. However, now that we have identified the cause as residing in the way we do things, the solution is to make certain changes in the way we teach and train people so that we can avoid the interference problem altogether.

The 'attitude problem' fallacy

When people don't follow instructions or procedures it is often because of accelerated forgetting. Admittedly, non-compliance may initially have started because of annoyance, distractions, fatigue, work pressures, poor training or other possible reasons. But if allowed to go on uncorrected, non-compliance (doing nothing or doing the wrong thing) gets to be repeated (practiced) and soon becomes a learned error and is then much harder to change.

Once this wrong behaviour is entrenched, it is maintained by force of habit *(automaticity)*. Importantly, while the person's 'bad attitude' may have initiated the unfortunate behaviour, 'attitude' is no longer what is maintaining that behaviour - proactive habit interference is now keeping it going.

The popular misbelief that continued non-compliance is mostly due to a 'bad attitude' leads to another popular fallacy, namely that 'employee education' will fix the problem of non-compliance.

...inhibition!

The 'employee education is what's needed' fallacy

Whenever we want a person to change the way they do things we try to 'educate' them. This process involves getting their commitment, cooperation and attention; explaining how important it is for them to change and what the benefits are; pointing out possible consequences of not changing; and developing their conceptual understanding of what is involved in the 'new' way of thinking.

This instructional progression is similar to the conventional teaching process used in almost all training and re-training programs described earlier and that is why it is designed to fail. Because the learner is not a blank slate, he already has his own, different, 'understanding' from previous training in different ways, different processes and procedures; from self-taught attempts; from observation of others; and from possibly incorrect or misinterpreted prior training.

Furthermore, the person's own understanding is protected from change by PI so the new 'education' suffers accelerated forgetting and he soon reverts to his old ways. Additional re-training or re-education attempts are unlikely to make much impact on this problem.

There is no denying that improving conceptual understanding is a key learning element in achieving behaviour change and improvement. What we are pointing out, however, is that conventional methods of re-educating and re-training, however well informed and well-intentioned, will suffer from the effects of proactive habit interference, as evidenced by trainee confusion and frustration, accelerated forgetting of the new knowledge and eventual reversion to old ways.

Accelerated forgetting and reversion to old ways are unfortunate side effects of the way we currently train people

What does all this mean for aviation training? To learn you have to be able to change your existing knowledge and skills. Most trainees are not a 'blank slate', they already have at least some knowledge or experience of the topic or skill being taught. Experienced practitioners come to the learning situation loaded with prior knowledge of the topic or skill being taught.

Change, necessarily then involves *unlearning* of prior knowledge and skills that may be incorrect or have become outdated. *Unlearning* is very difficult and slow because old habits and skills die hard, due to the brain's knowledge protection and maintenance mechanism. Conventional re-training and re-teaching methods inadvertently activate this proactive inhibitory effect which causes massive interference with the learning process, accelerated forgetting of the new knowledge and skills, repeated episodes of reversion to old ways, a lot of frustration and unnecessary expense, and a greatly prolonged time to achieve competence in the new skill. That is our dilemma as human learners. How can we overcome this fundamental problem?

There *are* special ways of dealing with learned errors or habit errors but almost all of these ways:

- require specialist knowledge
- are difficult to learn
- require close monitoring by the trainer

- rely on external rewards or punishments, thereby creating problems of supply and control and also leading to unfortunate side-effects
- are resource intensive and not cost-effective.

Even if they do not suffer from these problems these methods are usually slow to work. Clearly, we need a better way.

A new way of changing and improving

Of all the books, literature and various academic publications available on the subject of learning, perhaps the least used method in modern day airline training organizations, although argueably the most innovative and effective is the 'Old Way/New Way Method' *(© E.H. Lyndon 1973)*. This approach offers educators and learners a viable and cost-effective alternative to straightforward re-training and avoids the pitfalls associated with this approach.

Old Way/New Way is the result of many years of effective collaboration between researchers and practitioners in many applied fields. Consequently, it is not only an effective learning method but it also has a sound theoretical underpinning. This theoretical framework is a novel interpretation and synthesis of established learning principles and is soundly based on established research in learning, errors, habits, memory and transfer. This methodology has strong implications for the acceleration of learning in all areas including conceptual as well as skill learning.

...pick up missing knowledge!

Instead of always re-teaching, the trainer now has an additional methodology to add to his or her repertoire of skills. It is important to point out that Old Way/New Way is not a replacement for good teaching; it is a complement to effective teaching and is therefore a useful add-on that trainers can readily incorporate into their professional tool box. Which method to use, i.e., re-teaching or Old Way/New Way, depends on the results of relatively straightforward diagnostic testing and error analysis. Let's see how all this works in practice.

Old Way/New Way

When people don't do what they are supposed to do, when they fail to comply with operating procedures, when they appear to forget what they have learned in training and start to go back to old ways, these are clear signals that conventional training and re-training may not be working properly. The following steps should then be implemented:

- Check if the behaviour/performance is consistent or inconsistent.
- Consistency means the person does the same thing every time or almost every time, whenever they do it 'wrong' inconsistency means that the person does it wrong but each error is different or variable - there is no obvious pattern.
- If the performance is inconsistent then you can proceed with conventional re-training.

If the performance shows consistency, then:

- You are dealing with the presence, rather than the absence, of learning (i.e., the person *can* learn, after all). The person now has a learned error, i.e. a 'learned disability' rather than a 'learning disability' in that he knows how to do it 'wrong' (he takes the wrong action or fails to act). You have to respect that knowledge and start from there, rather than assume he knows nothing. So instead of just retraining, i.e. repeating the initial training, the solution is to use Old Way/New method.

When the original performance problem is correctly diagnosed, the Old Way/New Way correction protocol is properly and sensitively implemented and appropriate follow up procedures are put in place, then the likelihood of activating the mind's inbuilt knowledge protection and maintenance mechanism is much reduced. Proactive interference is therefore greatly reduced or is non-existent. This then avoids the problem of accelerated forgetting of the new way and the associated reversion to old ways so often encountered when using *conventional* training methods.

...to hear is to forget, to see is to remember, to do is to understand!

By the simple use of language, Old Way/New Way is able to reduce proactive habit interference, thus allowing unlearning of the old way to occur and the learning of the new way to proceed much faster and with more permanence.

Time and time again, trainers who have seen this method in action have been surprised, even amazed, by the results. However, the more observant of these have pointed out that before a trainer could incorporate and integrate this worthwhile method as part of his 'tools of trade', he would need to *change* his *own* training routine. As we all know, old (training) habits also die hard so this is no easy task. Fortunately, Old Way/New Way courses for trainers are specifically designed to overcome this problem.

What are the limitations of this method?

A problem with Old Way/New Way is that to an inexperienced observer it is deceptively easy. Educators and trainers are always surprised by the apparent simplicity of the technique and assume that if it is that simple then surely anyone can easily pick it up. However, like most procedures that are sophisticated but look deceptively simple in the hands of experienced practitioners, there are many traps for young players. Just like an iceberg, there is much of significance below the surface that lies hidden to inexperienced eyes. Prospective users of the methodology should complete an approved training program with post-training implementation support, otherwise they may short cut the method; they soon start to leave steps out, add extra things and change the instructions. This means that their initial successes soon start to fade. They then blame the method, start to mistakenly believe that it will not work for them and then stop using it.

Another requirement, which is sometimes problematic, is that Old Way/New Way requires a trainer who can truly fulfil the role of subject matter expert. More than with any other learning process, this methodology needs a practitioner who can conduct valid diagnostic tests and use the results to identify any Type 1 and Type 2 errors that may be present. In other words, the trainer has to know what the trainee is doing wrong, what he or she should be doing instead and precisely how these performances differ. This requires a 'thinking' trainer. Unfortunately, not every trainer is interested in errors and their correction to this extent. Old Way/New Way is of little use to such persons.

...a thousand times!

Years of experience tells us that Old Way/New Way is better able to achieve permanent behaviour change when the learner is an informed participant in the change process. While it is nevertheless possible to apply the methodology and 'do it' to a person and achieve change, the process is more likely to be successful in the long term and with a wider range of behaviours when the learner receives some preliminary instruction in the learning principles that underpin Old Way/New Way. This is entirely consistent with the premise that we cannot change people; they have to change themselves, i.e., it is the *learner* who has to use Old Way/New Way to change him/herself. It then becomes the role of the Old Way/New Way facilitator to 'teach' and assist the learner how to use Old Way/New Way to achieve the desired changes in understanding and in behaviour. This requires the facilitator to share his or her professional expertise with the learner or trainee, otherwise the learner cannot be empowered with Old Way/New Way. This transfer of knowledge and expertise and the resulting empowerment is a sophisticated and demanding process that not all trainers have the time nor inclination for. Clearly, Old Way/New Way is not for everyone.

Most of these prerequisites are certainly not unique to Old Way/New Way; it is just that they place limits on what can be done to enhance learning using this approach. However, in the hands of an expert trainer who is seriously interested in and committed to the development of human understanding and skill, who will take the time to closely observe and diagnose learning difficulties and scrutinise errors for consistency, and who is willing and able to share his or her expertise with trainees, Old Way/New Way is a most powerful learning tool that clearly represents best practice in continuous improvement and change management.

The author acknowledges Dr Paul Baxter's input to this chapter and the methodology of the Old Way/New Way process.[1] Dr Paul Baxter uses and teaches Old Way/New Way techniques to airline training organizations.

The process of learning is indeed a complex subject. The human brain and its memory is an incredible machine, capable of astounding feats. The next chapter looks a little deeper into how we learn and why we forget.

---

[1]Information on this fascinating subject may be found by writing to Personal Best Systems, P.O. Box 197, Mt. Ommaney, Queensland, Australia, 4074. Further information including all published research, case histories and details of online courses and consultancy services is available; see web site link in index.

Individual interactive computer based learning forms an essential part of aircraft systems study when converting to new aircraft types.

7 The brain - memory

Many people experience major problems in such areas as thinking, memory, concentration, motivation, organization of ideas, decision making and planning. Airline pilots are required to remember, and retain, a host of facts and figures, and organize their thinking processes along defined and structured lines. The aim of this chapter is to look at ways to improve memory and concentration skills.

In high-pressure situations, involving multiple task operations, the brain is robbed of its ability to choose or select on what to concentrate. Only learned reflexes and practised habits during embedded in the subconscious will dominate and control how you behave.

The problem with this type of decision and action is that much of it is automatic i.e. below the level of conscious awareness. It follows that if you previously learned incorrect or inappropriate techniques then you will most likely inadvertently revert under conditions of high stress or mental workload.

Task management

Experiments show there is an optimal level of workload, below which and above which, individual and total task performance degrades. Humans attend to tasks in such a way as to balance workload to achieve acceptable levels of performance in task management.

Recent lines of research focussing on task management rather than workload management have formulated the theory that 'human operators controlling complex systems engage in a process called task management. Task management is a high level mental process which continuously prioritizes concurrent tasks and allocates resources to them'.[1]

---

[1]'Fight Crew Task Management', Schutte, P.C., & Trujillo, A.C. (1996)

Preoccupation with one task to the detriment of the other is one of the more common forms of error. We perhaps all remember the more dramatic consequences of this type of error: the B707 that ran out of fuel whilst the crew were preoccupied with a landing gear malfunction; the DC-9 that landed with the gear up as the crew were trying to sort out an unstabilized approach; the L-1011 crash that occurred whilst the crew were preoccupied with a landing gear light problem and failed to notice that the autopilot had become disengaged.

Cognitive research indicates that people are able to perform two tasks concurrently only in limited circumstances, even if they are skilful in performing each task separately. One cognitive system involves conscious control whilst the other is an automatic system that operates largely outside conscious control. The conscious system requires effort, thought, is relatively slow and tends to perform one operation at a time, in sequence. Learning a new task requires conscious processing. Automated cognitive processes develop as we acquire skill; being task specific, they operate rapidly and fluidly, requiring little effort or attention.

A task requiring a high degree of conscious processing cannot be performed concurrently with other tasks without risking error, FMS programming for example. Even something routine like conversation can cause distraction or overload. Pilots obviously have to learn to integrate these tasks if they are to achieve reliable and satisfactory performance.

'Pilot error' in aircraft accidents/incidents, may be an error in *performing* a flightcrew function, but as statistics show, the error is more likely to be an error, or series of errors, in *managing* activities. Sound, flightdeck task management is the desired goal. As there are generally multiple, concurrent tasks to attend to, the crew must create an initial list of tasks to perform and then continually:

- Assess the current situation
- Activate new tasks in response to recent events
- Assess task status to determine if each task is being performed satisfactorily
- Terminate tasks with achieved or unachievable goals
- Assess task resource requirements
- Prioritize active tasks
- Allocate resources to tasks in order of priority
- Update task list.

How adults learn

- If they want to
- By linking past, present and future experiences
- By practising
- By help and guidance
- By being in an informal, non-threatening environment.

Brain structure

Neurologists are now saying that the average brain contains 100 billion brain cells. Each one is like a tiny tree with messages passing from branches to roots, each making hundreds of connections to other cells as we think. The total megabyte capacity is inconceivably large.

Ageing

If the brain is stimulated no matter at what age, new *twigs* will grow, each cell branches and increases the total number of possible connections. New brain cell connections can be generated provided the brain is stimulated and exercised.

No human yet exists who can use all the potential of his brain. The average person can think 800 words per minute, but the average trainer/lecturer can only talk at 120 per minute. Our trainees need something interesting to do with their spare 680 words per minute.

Dominance

Professor Roger Sperry, a 1960s Nobel prize winner, is credited with the findings that the two sides, or hemisphere, of the brain tend to divide major intellectual functions between left and right.

Left	*Right*
Speech	Artistic activity
Calculations	Musical ability - rhythm
Intellectual analysis	Emotions
Reading	Recognition

Writing	Comprehension
Naming	Perception
Ordering	Spatial ability
Sequencing	Facial expression
Complex motor functions	Holistic ability
Critique	Intuition
Evaluation	Creativity
Look	Images - colour

Although each hemisphere is dominant in certain activities, they are both skilled in all areas. The mental skills identified by Sperry are actually distributed *throughout* the brain.

It has been fashionable to label people either left or right dominant. This labelling may well be accurate but by inference it does limit and is a barrier to developing further expression and application of all our mental skills. Trainers should encourage trainees to use both sides of the brain. It has been shown that people who have been trained to use one side of the brain more than the other find it difficult to switch when necessary. When the weaker side is stimulated and encouraged to co-operate with the stronger side, there is a great increase in ability and effectiveness.

Psychology of learning

Research has shown that, during the learning process, the human brain primarily remembers:

- Items from the start of a learning period
- Items from the end of a learning period
- Associated items, patterns stored, or linked to other aspects
- Emphasized items, outstanding, unique
- Items which appeal to the five senses
- Items of particular interest.

Recall

When a message is given once, the brain remembers only ten per cent one year later. When given six times, recall rises to 90%. The message is: *repeat, recap and review.*

Retention

The use of words *and* pictures will create greater impact and a lasting impression, thus aiding retention.

...the capacity of recognition memory for pictures is almost limitless!

The astonishing machine

Your brain has five major functions:

- Receiving
- Holding
- Analysing
- Outputting
- Controlling.

These five categories all reinforce each other.

It is easier to *receive* data if you are interested and motivated. If the data is received efficiently, you will find it easier to *hold* and *analyse*. Efficient holding and analysis will increase your ability to *receive*.

    Information that has been efficiently received, held and analysed, will be easily *outputted* or expressed by speech or gesture. Controlling the *output* depends on the brains general state of health, your physical and mental state. *Healthy body = Healthy mind!*

Mind set

The enemy of recall is *mindset*. When people hear or see something that clashes with their beliefs or values they distort or simply reject it.

The power of images

In the 1970s, a series of experiments was undertaken in an attempt to show the amazing retention power of the brain.

Each person in the experiment was shown 2,560 photographic slides over a controlled period of several days. They were then tested for recognition, by being shown two sets of 2,560 slides, one set they had already seen, the other they had not. Their average accuracy of recognition was between 85 - 95%.

A second experiment was designed to check the brain's ability to recognize at speed. In this experiment, one slide was shown every second. The results were identical indicating that not only does the brain have an extraordinary capacity to imprint and recall, but it can do so with no loss of accuracy at incredibly high speed.

The third experiment consisted of showing reverse images, at the rate of one per second. Again, results were identical indicating that even at high speed, the brain can juggle images in three-dimensional space with no loss of efficiency. Further research found that if vivid striking, memorable images were used then accuracy rates improved towards 90%.

...a picture is worth a thousand words!

Images are more evocative than words, more precise and potent in triggering a wide range of associations, enhancing creative thinking and memory. This shows the great disadvantage of standard note taking and note making without the benefit of images.

In the modern world of communication, with the ease of creating and transmitting pictures, the balance between the written word and images is coming into focus. For one hundred years, the accent has definitely favoured the written word and this in itself has caused some stifling of the brains enormous power.

Note taking and note making

Standard linear note taking and note making, consists largely of words, symbols, linear patterning and analysis. By using only these *tools,* many of the items essential for overall brain function are missing.

This type of note taking/making tends to show an absence of visual rhythm, pattern, colour, image, visualization, dimension, spatial awareness, and association. These elements are of vital importance to assist memory recall during the learning process.

Taking notes can be frustrating and boring. If the brain is bored it turns off, and goes to sleep. To be effective trainers and trainees, we need to revise the way in which we present and record information.

...if the brain is bored!

Note taking/making - aid to memory

- Organize your note taking and note making in hierarchical fashion, rather than randomly
- Avoid using lists, they tend to generate an idea then cut it off without any association, stunting the natural thinking process
- Use emphasis, image, colour, dimension, spacing
- Use association, arrows, colour, codes
- Be clear, use simple key words
- Develop your personal style
- Use numerical order
- Break mental blocks, use blank lines, add images, ask questions
- Reinforce, review, do checks, repeat
- Prepare mental attitude, copy, make commitment.

A positive mental attitude unlocks the mind, increases the probability of making spontaneous connections, relaxes the body, improves perception and creates a general expectation of positive results. Note taking is not just a memory aid, other functions such as analysis and creativity are equally important. The best notes will not only help you remember and analyse information but act as a springboard for creative thought.

Acquiring a mental set *(for example when reading an aircraft flight manual)*

- Browse through quickly, getting a feel
- Work out length of time available
- Establish associated mental *grappling hooks*
- Define your aims and objectives
- Make an overview, contents, headings, conclusions, summaries, illustrations, graphs and other eye catching elements
- Preview
- In-view
- Review.

Intuition - superlogic

Contrary to widespread opinion, emotions are integral parts of any decision making process. Intuition is a much maligned mental skill. The brain uses *superlogic* in order to access its vast data bank, in relation to any decision it has to make. Studies have shown that managers, chairman and presidents of multinational organizations attribute 80% of their success to acting on intuition or *gut feel*.

...relying on intuition!

Retention

It has been found that material that has been thoroughly learned is highly resistant to forgetting. Experiment has shown that:

- Rote learning tends to be superficial and short term
- Meaningful learning, using as many of the laws and principles of learning as previously discussed, goes deep, because it involves principles and concepts anchored in the learners' own experience.

There are *five* basic principles of retention:

- Praise gives pleasurable experience, and positive response stimulates remembering
- Association, the *trigger* response, promotes recall.
- Favourable attitudes aid retention
- Learning with all our senses is most effective (much attempted learning is through eyes and ears only)
- Meaningful repetition aids recall.

Short-term memory is very useful but limited. *Long-term* memory is almost without limit. For example, experience has shown that in dialling an unfamiliar seven-digit telephone number, 40% of a given sample of people pause to refer to the written number again, before being able to finish the dialling. Even very rapid rote learning and repeat (a few seconds) of another unknown number, using the same group of people, reduced the percentage that paused to re-consult to 20%. Short-term memory can be increased very rapidly, but falls away just as rapidly.

Factors which cause forgetting

- Disuse of information that has been learnt. It would appear that the information is never really completely lost from the brain memory cells but that the difficulty lies in summoning it up to consciousness

- Interference. Two basic principles appear to apply here

  > Closely similar material seems to interfere with memory more than dissimilar material. For example attempting to learn Latin and Italian languages side by side. There is danger, from interference in flying similar, but different aircraft, within a closely spaced time scale.

  > Material not well learnt in the first place, is the first to suffer interference.

- Repression. Material which is unpleasant or producing anxiety is subconsciously repressed

- Distraction and loss of interest

- Interruption. It should be noted that short term memory is very vulnerable to the effects of fatigue, stress, drugs and hypoxia. Food digestion after a heavy meal will reduce the ability to remember. All of the above reduce the amount of oxygen available to the brain

- New facts and skills are best introduced in the morning, rather than the afternoon.

Memory and increasing age

Short term memory suffers with age, in capacity, and from interference effect in particular, for example more interference effect with increasing age. In training older people, it should be remembered that they progress better by doing and watching the success of their actions, rather than by giving them complicated instructions to remember. The learning of high-speed *motor skills* in particular falls off rapidly with age, for example, computer or complex mechanical operation.

Older people may progress better if the pace of training is under their own control. Furthermore, although slower in perception and reaction time, there is an offset effect brought about by their higher level of experience. They may reach a similar or better skill level than a younger person in a slightly longer time.

Learning lists, facts, sequences and drills

There is a requirement for aircrew to learn short lists of facts, emergency drills and other sequential operations. Let us look at some of the techniques involved and used by the advertising industry, as an analogy of a good trainer, of using effective methods to help his trainee learn some facts. These techniques are all legitimate methods in the theory of learning. An exception is subliminal television advertising, which is banned in many countries as it approximates to *brain washing* of the subconscious mind.

It can be said that we retain about 10% of what we hear and about 30% of what we see and hear (short term). If the product has been demonstrated and used, we retain about 80% of the relative facts in what we see, hear, and do, as well as what we think about while we are doing it. For example, in the case of a test drive, in a new car, if the salesman appeals to our sense of reasoning, which can be defined as our capacity to discern similarities and arrange facts, he is well on the way to a sale, even before he starts to appeal to the more basic instincts of ego or acquisition.

The following is a list, by no means complete, of some of the methods used for short-term recall, and enhancement of a learning situation. With imagination, many of these methods are suitable for your use:

- Chants and rhythms - children learning multiplication tables
- Jingles, songs, rhymes - TV advertisements
- Catch words - slogans
- Colour and highlighting - enhances any learning situation
- Arrangement and set out- good text books and manuals use wide   margins,   white   spaces, effective sub-headings
- Super sell, saturation - *block-buster* movies and books
- Dynamic motivation - use of sex in advertising, fast cars, learning by association and appeal to prime instincts

- Abbreviations/acronyms - NASA much easier to remember than National Aeronautics and Space Administration
- Key words, letters, - Learning by association, easier to phrases, number groups remember than full words, can be arranged into a pattern easily.

Mnemonics

- Activate brain cells on all levels making it more alert and skilful at remembering
- Intrinsically designed to aid memory
- Encourage probability of spontaneous recall
- Each usage increases base memory skills
- Enhance creative thinking skills
- Maintain high level of recall
- Utilize the individual's association capabilities.

Mnemonics are fun to make up, and are very effective for both short term recall and long term recall. If the letters can be kept in order of importance and still form meaningful memory words they are even more effective.

Brain storming

Another way to help you remember and expand ideas is to use the 'mind mapping' technique. Starting with a clean sheet of paper, write the aim in the centre. Draw a circle around the aim and then by drawing lines away from the centre write a key point or word on each one. From each keyword line, subdivisions can be made of the important points relevant to each key point. Further subdivisions are then possible. Use colour to emphasize different areas and let your mind run wild.

Once all the ideas are written down then the diagram can be edited appropriately and re-drawn. This technique can be particularly effective, for example giving a briefing, as it can be adapted and made interactive. Each time you see or hear a keyword your memory will be jogged and you will pick up the next thread. The main drawback to this system is to ensure that you do not ramble on, or go off at a tangent, overrunning your allotted time.

As an tool for creative, lateral thinking 'mind mapping' can be applied to many types of learning and teaching situations, opening exciting and stimulating possibilities. Visually more eye catching than standard linear note taking whilst adding a new dimension to expanding the memory.

Mind map

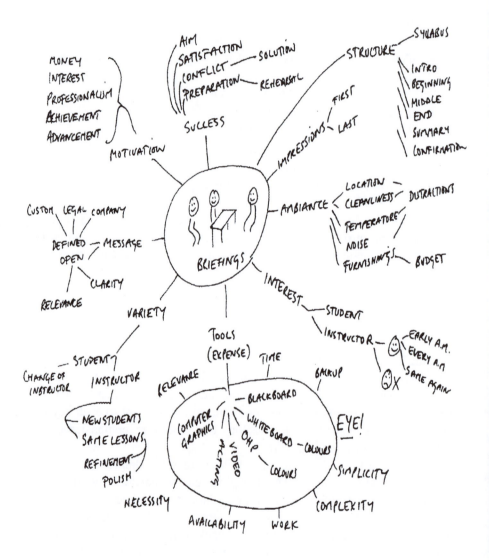

What can we learn from mistakes?

Good habits and skills allow us to function effectively. However, when our actions and reactions are inappropriate then the outcome is undesirable or even downright dangerous. Many errors result from careless mistakes, distraction, inexperience, inattention, or just lack of motivation. Practice and application can overcome these kinds of errors. However, errors that are more deep rooted are a sign that the person has learned something but learned to do it wrong i.e. habit error. Error analysis can help identify which kind of error is present by identifying the pattern. This 'habit error' is more of a problem to eradicate as in involves unlearning what already is embedded in a person's subconscious – 'Old habits die hard'.

(see Chapter 6, The process of learning 'Old Way/New Way' on how to remedy this.)

... let's use dynamic motivation!

We can learn a great deal from other people's mistakes, for example, studying accident/incident reports, researching further our own errors and infringements, finding out more about the unusual quirks of our own particular aircraft and so on. Human error is inevitable and inescapable and can never be *completely* eliminated. However, we can reduce the likelihood of error, trap errors before they have operational effect and mitigate the consequences of error by sound error management principles.

The application of good standard operating procedures (SOP's) will go a long way to reduce errors. Trainers should ensure that their trainees are aware of human limitations and in particular draw attention to:

- Schedule activities to minimise conflicts
- Recognize that 'head down' tasks reduce the ability to monitor
- Recognize that conversation is a powerful distracter
- Importance of the scan, where two or more tasks are being performed concurrently
- Vulnerability of interruptions
- Ensure that pilot flying (PF) and pilot not flying (PNF) duties are explicitly assigned.

The complete mind

Leonardo de Vinci's principles for the development of the complete mind consist of the following:

- Study the science of art
- Study the art of science
- Develop your senses, especially how to see
- Realize that everything connects to everything else.

Genius is the capacity to see ten things when the ordinary man sees one.

It is perhaps an appropriate moment to move on to the practical application of how perception and personal skills are incorporated in everyday flight operations. The next chapter looks at some of current airline training practices and operational problems that confront the average line pilot.

Clear concise diagrams of complex electronic flight information systems form a useful training aid to learning and understanding.

Airbus A321 Flight Simulator – Lufthansa Flight Training – Frankfurt.

8 Training objectives and methods

This chapter comprises notes compiled from many sources and reflects a wide experience. Aimed primarily at explaining current training policies and techniques, it is hoped to offer some guidance to new pilots embarking on an airline career. They cannot embrace all practical thoughts on training and should be considered as a limited information resource and as a catalyst, to encourage your own personal research, and development of your own personal skills. These notes are intended to assist in the *how to* train area, and as such, must always be considered as advice, subject to the overriding status of the operations manual and other authorized documents. Trainers should refer to the line training form, and the trainee's own training file, for items of training that require signing off, by the training pilot.

Training objectives

As a training pilot, you need to have a clear perception of what you are seeking to achieve, related to the experience and rank of your trainee. You have the creative task of providing the opportunity and encouragement for a pilot with known skills and knowledge to become proficient in his new operating environment.

Professionals performing poorly

Pilots not only make mistakes but also often break the rules. So say the Royal Aeronautical Society (RAeS) at a recent seminar in 1998. This statement is backed up by statistics from NASA Ames Research Center and findings by Boeing who analysed airline hull loss accidents over a period of ten years. They concluded that many of the crashes studied could have been prevented by pilot adherence to published procedures. More so than any other 'accident prevention strategy'. Training has to do more than impart knowledge and skills; it has to create a level of consciousness, which enables a pilot to recognize problems met in the real world. Trainers should be particularly vigilant in pointing out non-standard procedures when checking or evaluating routine training sessions.

Standard Operating Procedures (SOP's)

Adherence to SOP's are the backbone for all training/checking/operating sessions whether initial, type training, or recurrent and must be clearly stated. Problems identified with adherence to SOP's that would require addressing by the trainer:

- Ignorance of the procedure
- Tiredness and fatigue
- Time pressure, commercial schedules, slots, duty limitations etc.
- Poorly written and ambiguous manuals
- Belief that departing from a SOP is justified either because of circumstances or because the procedure seems wrong or unnecessary
- A 'can do anything, I know best' mentality
- Maybe there are too many SOP's that are not relevant or need updating.

The line check

The line check is considered as the *gateway* through which the trainee should pass, with reserves of skill and capacity. The successful line check means that the pilot has attained a standard of competency to permit his continuing development on the line. Pilots should be seeking to develop higher standards of excellence throughout their career.

Manipulative skills

Personal standards should be the highest possible, to ensure adequate margin over the aviation authority and company standards. It is in this area that line crews find the greatest satisfaction and professionalism. In general, the standards required by legislative air navigation orders, for the maintenance of an airline transport pilot licence (ATPL), and instrument rating (IR), should be regarded as both a *starting point* and a minimum for every pilot.

There are three levels of manipulative skills:

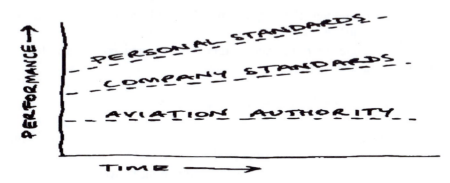

Training plan

It is important that the trainee is provided with a clear idea of the training plan. It should be progressively explained how it will operate, throughout the training period. The training pilot should explain that he will use the plan flexibly, so as to benefit from operational situations as they occur on the line. The training pilot should provide feedback on the plan, as training progresses, to assist him in developing it further.

Study guidance

The trainee will need guidance on his priorities in learning. Here the trainer can help him to keep his studies pertinent to the training plan and help him to organize the operational data he needs. For example: noise abatement runways, reduced thrust policies, special take-off procedures, CAT 1, 2, and 3 runways, reduced visibility takeoff.

Provide an overview of the phases of training

The trainer should explain at the outset, and again as the training develops, the phase of training which is currently in use.

Maximum assistance phase

In the early stages provide full support and help your trainee to settle in. Make good use of the safety first officer. He can assist the trainee pilot with external and internal checks, fuelling, documentation, correct R/T terminology, and generally organizing his workload. The safety F/O should be carried until the trainer is confident in the trainee's ability to land the aircraft, operate the radio, and demonstrate a sound knowledge of memory items and limitations.

Gradual withdrawal of assistance phase

Ensure that the trainee understands that this phase is normal and is introduced in order to increase overall awareness. As he gains competence in normal crew duties, the trainer should progressively and sensitively withdraw his assistance to guard against training pilot dependency. Never allow the performance or security of the aircraft to be degraded during this phase. Knowingly or unknowingly, the trainer may be providing verbal or non-verbal cues, such as a hand next to the gear or flap lever, being too early with backup calls, prompting the checklist or suggesting fuel considerations for example.

There are two important guidelines to consider in this phase:

Allow your trainee to make mistakes and discover his limitations. We train for systems failures, we must also train to detect and prevent human failures. This must always be handled positively without destructive criticism. Do not intervene too early. Allow time for your trainee to discover something has been overlooked. As training pilot, it is your responsibility to ensure safety is not compromised at any stage in training. It is all too easy to become blinkered and distracted.

Normal crew operations phase

When the trainee can operate the aircraft in his own right, the trainer should move the training into the next phase of a relaxed competent normal operation, with good crew co-ordination in preparation for the line check.

Well I'm sure you will check the fuel before we leave next time

...allow the trainee to make mistakes!

Pre-flight briefing

Initially, the trainer should provide, or draw out, a normal flight profile and discuss each phase of the flight for normal operations. Ensure descent profile techniques are understood. Remember that *a picture is worth a thousand words*. Later in training, ensure the trainee clearly understands any specific training tasks for the flight. Give him time to review and assimilate, to think and plan. In this way, he will best learn from his experience.

Pre take-off briefing

Provide and/or discuss what is required for take-off briefings.

Use of automatics

The trainer should explain how he will introduce training on new technology. Discuss its limitations and potential for distraction, such as re-programming a tracking change during a standard instrument departure (SID). Here stress the requirement for sound flight management practices. There should be a sensible blend of manual and automatic flight, concentrating only on those areas that are demonstrated to require attention.

Use of training aids

The training pilot will need to spend time with his trainee. Use an aircraft on the tarmac or company learning aids to discuss for example drills and procedures. If a *training plateau* is reached, it may be appropriate to refer the trainee to the training school for further specific instruction or use of *hands on* simulators if necessary. In this way, you can help to separate the pressure of flying the aircraft, from the learning of new skills. Remember the trainee will undoubtedly be under a great deal of pressure in his new environment, and a sympathetic understanding of his problems will greatly enhance the training environment.

Time management

Arrive early before a flight in the first part of the training. This allows time for discussion and planning the day's flying. The trainee will be encouraged by the trainer's dedication. Some new trainees will come under significant pressure of on-time departures and may need assistance to manage the pressure of time, on the basis that preventing operational delay is the top priority. As a trainer, do not overlook the help and advice for the items often taken for granted such as car parking, office layout, location of briefing room, company paperwork.

Debriefing

Always try to review the experience, as near to the time it occurs, as possible. Much of the actual learning process will occur whilst in the aircraft, and there are excellent reasons for this. People can recall data much more efficiently if they are in the environment where they learnt the material. It follows if a person is expected to operate in an aircraft his learning should take place there. However, there will be occasions when an empty briefing room or a hotel room on an overnight will provide a more suitable learning situation, so use the facilities that are available.

In a straight lecture situation, a person's attention level drops significantly after about twenty minutes. The attention span is increased if a person can perform a related activity following the talk, and their span is further increased, if the trainer then summarizes the material. So when planning, for example, to go through the operations manual, plan breaks, use diagrams etc. in the analysis. It is unproductive to keep the student for too long when he is tired, or it is late in the day.

...under some pressure!

Feedback

Invite, and seek from the trainee, positive or negative comments about *how it's going.* Specific feedback may resolve difficulties and avoid brooding or premature involvement of a third party.

Use of training forms

Allow the trainee to see any training forms that apply to him, and let him, if possible, read your comments before they are submitted. This will avoid suspicion or doubt about your integrity as a training pilot, and thus build confidence. Try to keep your comments related to what has been achieved, rather than each specific difficulty. If you need guidance in a training situation, seek it out privately from the appropriate flight training department manager. Negative comments on training reports should be confined to problems which are not easily rectified and which you believe warrant flight department awareness.

Explain your role as training pilot

Emphasize your role as one of encouragement and assistance. The object is to make the training as enjoyable as possible and to see the potential. This will be reinforced by your conscious use of:

...attention level drops!

Motivation

Avoid the use of anger/fear/cynicism/negative criticism and discouragement. Appeal to airmanship, encourage and phrase things positively. Give recognition wherever possible, a little praise goes a long way. Your freshness of outlook and enthusiasm will encourage the trainee.

Analysis

At the completion of a task, ask the trainee for his own appraisal of his work. Use positive type questions to draw out other aspects. Then provide your own insights with honesty and sensitivity.

Understanding

Where a difficulty arises, try to identify with the situation. Remember your own training problems! Try and see the problem from the other person's point of view. How could you best be helped if you were in his shoes? Be aware of your trainee's progress and introduce new aspects at the most appropriate time. Make allowances when appropriate, and do not try to remedy every situation at once. Do not demand too much too soon. Assess the priorities and seek to resolve one issue at a time. This will avoid overloading the trainee and result in the quickest progress.

Effective communication

Choose words, which convey meaning, for example: the trainee is not doing well during an instrument approach. You keep on saying *come on, stay with it, keep your scan going.* He tries to comply but does not improve. What you really should be saying is *you're fixing your attention on the horizon indicator* or *keep a constant angle of bank* or *hold the correct pitch*. Help him to understand exactly what you mean, and what specifically is needed for improvement. In this case, you could review the operations manual on instrument flight, once on the ground. Relate your degree of verbal instruction to the individual. Some trainees will need a lot of explanation, others will function best with a minimum of verbosity. Consciously assess what is needed in this respect.

...giving encouragement and assistance!

Flight planning and in-flight re-planning - meteorological information

The trainee must be fully able to interpret and apply appropriate meteorological documents. This may, on occasions mean a personal appraisal of weather trends, where the training pilot suspects a change. Ensure weather forecasts are understood in concept. Encourage calculation of wind components where en-route wind gradient can affect the selection of flight level. Research incident/accident circumstances to learn from other's experiences of unscheduled diversions and weather related problems.

Ensure the trainee has a good understanding of the weather conditions conducive to the various types of windshear, and its operational consequences. The training pilot's local knowledge of likely weather patterns is an essential part of line training.

Fuel policy

The trainee should not only be conversant with company fuel policy, but also be able to exercise command discretion, and be given the opportunity of deciding on fuel load. This is an area of continuous balance between safety and economy. At the completion of fuel calculations ensure the trainee carries out a logical overview of the fuel components applicable to the flight, and that the total fuel required relates mathematically to the endurance in minutes.

Specific fuel considerations and questions

The following examples are the kind of fuel considerations, which the trainee should be considering:

- How much margin fuel will provide for an emergency diversion in case of unexpected runway closure? Consider particular single runway airports, and use reduction of reserves to emergency values.

- Anticipation of both improving and deteriorating changes to existing weather.

- When an alternate is required, how much additional holding fuel is desirable to avoid premature diversion?

- Additional fuel in some circumstances may conserve fuel by preventing an otherwise avoidable diversion.

- What are the fuel components to consider, fuel from the holding fix to the destination, instrument approach and go-around and engine out or depressurized diversion?

- What are the special considerations for the use of cruise level diversion route?

- When holding for a closed destination, divert fuel may have to be based on descent to destination, overshoot and diversion.

- In calculating alternate fuel for an airport where fuel is not available, consideration will have to be given to the fuel required to descend and land, taxi in, and standard block fuel for the following sector.

- The trainee should be able to calculate diversion fuel in one or two minutes. Encourage him to logically overview his calculations and seek a crosscheck from the other pilot.

- With fuel tankering, has the trainee considered the trade-off between economy of fuel pricing and the effect on safety of take-off? At a lower weight there may have been runway surplus to requirements, which may be desirable to retain, in circumstances such as contaminated surface or windshear.

- Can the trainee correctly apply critical point (CP) and point of no return (PNR) criteria?

Take off and departure

In the early stages of training ensure sufficient time is allowed for an unhurried approach to pre-departure tasks. Later, the trainee will have to manage his time to ensure his duties are performed without short cuts, such as inadequate attention to TAFS and NOTAMS.

...anticipating the weather!

Aircraft energy management concept

From take off to shutdown encourage the concept of energy management, for example, potential energy and kinetic energy as controlled by thrust and drag, lift and weight. The management of these forces determines efficiency of operation. Take off from limiting runways is the most energy critical phase of flight, requiring sound understanding of aircraft performance and company operating procedures.

Take-off briefing considerations

The trainer should develop, and provide for the trainee, a take-off briefing that relates company procedures appropriately to the circumstances. The following typical take off considerations are presented as an example:

- Reduced thrust criteria
- Noise abatement procedure
- Intersection t/o criteria
- Specific t/o procedure (SID)
- 2nd segment climb
- Circuit direction
- Configuration and manoeuvre speed
- Minimum safety altitude
- EMC check list items.

Full briefing additional items

- Rejected take off procedure
- Engine out climb priorities
    - gear/flap selection
    - flight path monitoring
    - a/c configuration changes req.
    - failure identification
    - emergency check list memory items
    - normal checklist
    - transponder.

Additional items to consider if night, instrument meteorological conditions, wet runway, thunderstorms, adverse weather, such as:

- Reduced visibility t/o criteria
- Wet/contaminated runway procedures
- Windshear procedures
- Radar use - cb location etc.
- T/o procedures for IMC obstacle clearance
- Minima for t/o and landing
- Diversion- altn- fuel- MSA
- Contingency fuel.

Recognizing failures

Determine that the trainee understands that engine failure must be indicated by two or more instruments. Discussion of previous rejected take-off accident circumstances will help to guard against aborting for spurious reasons. Ensure primary thrust indications on all take-offs are monitored by other instrument indications such as needles in the green bands.

Take-off

If adverse take off conditions do not occur during the line training, ask the trainee to consider a particular circumstance for the take-off and what he would plan for the contingency. For example, thunderstorms and microburst; meteorology conditions indicative of windshear; wet runways; IMC engine out diversion. How would these factors affect his operating procedures? Ensure correct use of radar pre-take-off use when CB cells are in the area. As the aircraft is lining up on the runway, ask the trainee to observe the overrun area for the reciprocal runway direction, so that he builds up local knowledge of possible local hazards.

Initial climb

Is the trainee using the correct scan? Primary attention to ADI for body angle, frequent cross checks of ASI, HDG and VSI to ensure no sink-out. Crosscheck of the standby artificial horizon (AH) and turbine/exhaust gas temperatures (TGTs/EGT's) during each initial climb.

...never be distracted!

Ensure positive rate of climb is being monitored on altimeter (ALT) and vertical speed indicator (VSI) before calling for gear up, and determine later in training, whether gear red lights and flap indicators are being monitored. Ensure the trainee knows what he is monitoring and is kept fully *in the loop.*

As a training pilot, consider the following in the training for this flight phase:

- The history and causes of standard instrument departures (SID) infringements and how to guard against this type of occurrence; de-programming a rehearsed SID following its cancellation in the take-off clearance; non-operational chat during taxi out etc.

- Ensure the trainee understands the inviolate priority, that *the nominated pilot must never allow himself to be distracted from monitoring the flight path.*

- Stress the need to be alert for any performance loss and to take immediate remedial action. Classic example was the B737 accident, where the aircraft on take-off, overran into a river, in icing conditions with engine anti-ice off. This was a result of only partial thrust being set, due to blockage of the engine pressure ratio (EPR) sensors.

- One of the few jobs that require multi-million dollar decisions, to be made in a split second, is the case of a malfunction during the ground run take off phase.

Established in the climb

Ensure the trainee is correctly and accurately trimming the aircraft before and after autopilot engagement. This may sometimes be an opportune time to introduce some brief questions and discussion. Firstly, discuss any operational aspect relating to the flight so far. Then introduce any training discussion you consider suited to the moment. Appropriate steps should be taken so that someone is *minding the shop* at all times. Encourage him to politely request deferral of discussion, if it is impinging on the operation of the aircraft.

Cruise

Use the cruise segment to relax the trainee. If he is under excessive tension, ease off or ask questions that are more obvious, so that he can be encouraged by getting right answers. On some occasions during training, break the intensity by leaving aside the training emphasis and just enjoy a line flight followed by a visual approach. This could be done for a number of sectors, particularly if the trainee is on a *plateau*.

- *En-route training*
  Use this time to cover navigational and other aspects such as:

- Emergency en-route airfields

- Using NDB tracking if the RMI is not being adequately scanned

- Check CP, PNR and diversion calculation capabilities. Use of ground speed and monitoring wind gradient for change of flight level to improve economy and/or comfort. The following is a CP rule of thumb:

$$CP = \tfrac{1}{2} \ way \pm wind \ correction \ (further \ out \ into \ headwind)$$

$$Correction: Over \ 500 \ nm \ use \ \tfrac{1}{2} \ wind \ component \ in \ nm$$
$$Over \ 1000 \ nm \ use \ all \ wind \ component \ in \ nm$$

Use of radar for ground mapping. Use of 1 in 60 rule for weather avoidance.

ETA management

Subject to company policy for individual aircraft types, cruising airspeed may be able to be reduced towards economy operation, where projected tarmac time is ahead of schedule. With jet aircraft, tarmac times can be estimated using cruising groundspeed over total distance to destination, plus 8 minutes. Speed reduction of 0.01 mach increases flight time by about one minute over distances up to 600 nm. Monitor groundspeed to maintain ETA, within operating limits.

Rapid descent briefing

The trainee should not *assume* normal destination, as unexpected airframe, engine or fuel problem, could result in unexpected diversion. Consider possible en-route alternates, at all stages of the flight. Ask the question, *what if...?*

Descent

Ensure the trainee develops the habit of checking the descent clearance, and other traffic, prior to actually commencing the descent. It may help to verbalize when commencing descent, *we're cleared to FL...* Outside controlled airspace it is important to nominate the descent point early to give the air traffic control flight service unit time to determine accurately possible conflicting traffic.

Energy management/descent profile

From experience, it is apparent that many pilots have difficulty in assessing their descent profile below FL150. You should be able to provide your trainee with a simple and accurate method to determine the drag or thrust required. The following method is one way this can be achieved: Determine the descent capability of your aircraft in still air using speedbrakes at the normal descent speed. For example B737 = 2 nm/1000ft descent.

Preparing for the approach

The precise techniques used for approach and landing depend, of course, on individual aircraft types, and information on this is will be extensively covered in aircraft operating manual. The following points emphasize particular training aspects. Remember that this is an area where you, as training pilot, are on top and relaxed, whilst the trainee may be under pressure and operating on his limits. Consciously recognize how much he is under pressure and adapt to his needs.

Has the trainee thought through all the applicable considerations? After a long flight, it is particularly important to mentally *freshen up* to be ready for the landing.

- Instrument approach briefing

- Divert fuel

- Landing distance appraisal, wet/dry runway and crosswind limits

- Meteorological conditions conducive to windshear

- Most appropriate flap setting and runway direction

- Circling criteria

- Go-around procedure, mental rehearsal, and diversion planning

- Taxiing and gate information.

Landing distance appraisal

Ensure the trainee develops an understanding of landing distances, for different weights based on MSA, ISA°, zero w/v, level dry r/w, landing flap. Then discuss distance and percentage variations for alternate flaps, wet runways, unserviceable antiskids, manual speedbrake and corrections for wind, temperature, and pressure altitude. This kind of research will be most beneficial in understanding aircraft landing performance, as related to the runways you use.

Wet runways

The effects of standing water on touchdown, crosswind technique, directional control and deceleration need to be discussed and understood.

Refer the trainee to:

- The aircraft type manual

- The aircraft type study guide

- Company publications dealing with crosswind and wet runway techniques, winter operations, accident reports, aeronautical information circulars.

Discuss the different types of runway surface:

> Standard bituminous concrete
> Seal coat
> Grooving
> Porous surface.

Discuss the choice of runway and flap settings:

> Aquaplaning on landing. *Formulae* to apply:
> Wheels spun up 7.7 x square root of tyre press.
> Wheels spun down 9 x square root of tyre press.
> *Where P = tyre press in Ibs sq. in.*
>
> *Metric: Spun up: 3.43 Kps.*
> *   Spun down: 2.93 Kps   (kilopascals)*

Crosswind landing considerations

Again refer the trainee to all the relevant publications, for the aircraft type, in order to establish an accurate understanding of correct technique. Crosswind effect on circuit flying should be thoroughly discussed. Drift correction, turning radius, headwind, or tailwind on base leg, timing and accurate flying. One of the most common errors is inadequate drift correction on late final approach. Pilots initially want to point the nose of the aircraft down the runway, resulting in poor landings, heavy side loads and directional control problems. Another potential problem is excessive *wing down* prior to touchdown. Trainees should if anything be encouraged to position slightly upwind on approach rather than downwind. The landing of course must be on the runway centreline.

General cues

Many trainees tend to ignore visual cues, which were basic to their early training, and rely too heavily on the artificial cues provided for them. For instance, the natural horizon seems to lose its significance in a simple visual exercise, such as keeping the aircraft on a straight heading. Another example is intercepting the runway centreline solely by reference to the localizer course bar. This frequently results in an overshoot of the centreline, when simple visual reference would have ensured an accurate interception.

Becoming visual from an instrument approach

Ensure trainee is familiar with the danger of *duck under* when flying part visual and part instruments. Standard back-up calls and instrument monitor down to declared minima are essential. Make a point of developing these calls as a natural action, from the monitoring pilot. Point out that it may not be possible to determine sink rate from visual reference to approach lighting until it is too late. High sink rates can develop without visual detection, especially in poor visibility.

It is not easy to determine visually, whether a reduction in the number of approach lighting rows visible, is caused by a reduction in visibility, or an inadvertent increase in aircraft body angle. Thus, glideslope or PAPIS protection must be maintained in marginal conditions as long as possible.

Research approach accidents so that you can emphasize the importance of approach vigilance, and benefit from the experience of others.

Discuss with the trainee the visual cues required to continue various categories of precision and non-precision approaches.

Black hole approaches / bad weather circuits

The black hole approach, particularly with no precision approach aid or slope guidance is, without doubt, one of the most difficult of flying manoeuvres. Constant vigilance, flying an accurate circuit and correct timing are most important. Good support calls are essential, particularly during a bad weather circuit. Point out that the wind in the circuit may be quite different from that given by the windsock or ATIS and the aircraft should be positioned in order to fly a reasonably long downwind leg, if possible, to assess and compensate for drift.

Once in the circuit, the go-round to the missed approach heading can usually be safely executed by initially turning towards the runway. The trainee needs to appreciate that in this situation, the vertical speed indicator (VSI) is the best insurance. Prolonged high sink rates are extremely dangerous. Accidents do not normally occur at sink rates around 500 to 700 fpm, if other criteria are reasonable. Even if low on slope due to visual illusion, low sink rates should provide sufficient time to recover, when runway, and subsequently landing lights, become effective.

Visual illusions

Ensure the trainee is aware of the *tricks* that can be played by the many visual illusions that exist on approaches carried out at night, in crosswinds, in rain or snow, or poor visibility.

Anticipating go-around

It is quite a common occurrence for trainees new to type to persist with approaches that are unstable, and, if continued, will place the aircraft in an unacceptable position. This is due to a number of factors. The primary ones being apprehension, lack of understanding of the aircraft performance, and a reluctance to admit that the approach has become unacceptable. Point out that a go around is normal procedure from an unsatisfactory approach, which is outside company limits, or tolerances.

The trainee should know his personal flying limits from which a go around should be conducted. The training pilot must emphasize company policy to go around as soon as an unacceptable approach situation is occurring or about to occur, and to announce his intention. This keeps the crew in the loop, and aware that the flying pilot has things in hand. Everyone must be mentally alert for a go around.

Descent profile leading up to an instrument approach

Early in the trainee's experience, allow him to be a little conservative on speed, preparatory to final approach. Similarly in more demanding instrument conditions, it may be better to be conservative on profile, and well prepared for the approach, than risk being fast and high with the possibility of a missed approach. However, it has to be remembered that in many busy airport environments, speed profile control is carefully regulated.

Using raw data information

A number of trainee pilots quite often fly the ILS, as a slave to the course bar and glideslope. This results in continuous chasing of heading and sink rate. For raw data ILS work or non-precision approaches, teach the trainee to fly the trends by trial and error sampling. Fly a heading to allow for wind effect and make small adjustments to this heading. Fly a known body attitude and constant sink rate. Make small adjustments to sink rate.

Energy management on approach

Excessive use of automatics will depreciate basic piloting skills. Automatics in training should be used in relationship to the manual skills of the trainee. Sufficient practice should be given in both manual skills and application of automation. Elevator controls glide slope - power controls speed.

Manual thrust levers

The trainee should know base line power settings to fly a stable 3° slope for the appropriate flap settings in typical conditions. Teach anticipation of base power settings when decelerating and then fine-tuning thrust for speed when stable.

Windshear on approach

Much information has already been written on windshear procedures. Without duplicating information already available, two wind shear conditions need to be highlighted in training. All other references in company manuals should be drawn to the trainee's attention.

- Downburst / microburst

- Performance decreasing shear

Sink-out on late final

The most common type of windshear encountered is rapid loss of headwind component approaching the runway threshold. The rate of loss of airspeed can be far greater than the acceleration capability of the aircraft. The method to combat sink-out is to pitch up the aircraft and accept excess speed trade-off for lift. However, this manoeuvre also increases aircraft drag. Therefore, elevator control must be combined with rapid thrust increase, commensurate with effect of sink-out. Energy loss through loss of headwind can most rapidly be compensated by increased energy from more thrust. The trainee must realize that the large thrust increase will result in aircraft acceleration once the windshear has been compensated. At this point, the effects of large thrust changes must be anticipated. *If in any doubt then go around.*

Do not let the trainee spend too much time with his head in the cockpit, once over the airfield boundary. If all parameters were ok crossing the fence, any windshear effects will be seen visually before they can be read on instruments. Refer to aircraft type and company operations manual.

...don't keep your head in the cockpit!

VASIS/PAPIS

The trainee should be taught that, while an ILS glideslope provides an exact threshold crossing height, and converges to a specific point on the runway irrespective of displacement from *on slope* indication, the VASIS, when flown with displacement, parallels the glideslope indication, but continues the displacement all the way to the runway. This can result in a high threshold crossing height and landing well down the runway, or worse, in the low case, landing short with a possible undershoot. Ensure that the trainee is aware of the fact that the increased azimuth cover at night permits the PAPIS to be visible on base leg, where clearance from obstacles is not guaranteed. He should not, therefore, rely on the PAPIS for approach slope guidance until aligned with the runway.

Flaring at night

Point out that, on the approach, the trainee should utilize all the approach aids including the radio altimeter. It is imperative that correct sink rates and speed (as described in the type-operating manual) are established and maintained early in the approach.

When established on a constant slope, the distance between the individual runway lights will remain constant. Maintaining this constant picture is an important concept of night approaches, and you should make certain that he recognizes deviations from slope (visible by the distance between the lights opening or closing) early in his training. Make a point of showing this to him.

The flare initiation height will be indicated first, when, looking towards the far end of the runway he will notice the light beginning to close up. Shifting his focal point back towards the approach end, he will notice the lights beginning to open up, giving him a feeling of sinking into the runway. He should now begin to raise the nose of the aircraft towards the landing attitude and simultaneously begin to smoothly close the throttles. The rate of change will depend upon the sink rate and local conditions.

Now is the time that he should change his focal point to the normal position and make minor adjustments to the aircraft attitude as necessary. He should resist the tendency to stare into the area illuminated by the landing lights.

Remember that the descent should never be stopped completely, particularly at night. If in doubt as to the height above the runway, he should resist over flaring. Merely flare the aircraft initially and hold that attitude. The touchdown will be within normal limits and far more acceptable than floating down the runway a few feet up. This leads to tail strikes, heavy touchdowns and over runs.

The initial apprehension caused by black hole approaches and night landings will soon be overcome, as your student begins to feel the strong sense of satisfaction that comes from well-flown approaches and landings at night. The following hints may be helpful in correcting a particular landing difficulty.

Aiming point

The aiming point should relate to the glideslope or PAPIS reference point until late in approach. Below 200 feet the runway marking on the threshold side of the GS/PAPIS reference point should be used, or as specified in the operations manual. However, the trainee should appreciate that the touchdown point will be a little beyond the aiming point depending on the trajectory of the flare. The acceptable touchdown area is as stated in the operations manual (normally the 1000ft (300m) point).

Flaring high/low

Your trainee's flare height judgement may be affected by the following:

- Runway slope
  It is common to flare low for an upsloping runway, adjust flare height in anticipation from local knowledge.

- Seat position
  Re-assess seat height if a flare problem is exhibited over a number of landings.

- Focal point
  Flaring low, focal point too close, flaring high, focal point too far. Encourage the trainee to adjust his focal point to compensate. Do not be surprised if initial over-compensation results. Some pilots find conscious eye movement to the end of the runway and back to the aiming point, is helpful prior to the flare.

- Night landings
  There is a natural tendency to flare high at night. A little anticipation will compensate.

- Thrust retard
  Late thrust retard may result in a tendency to float. Reduction of elevator back pressure is necessary to touchdown when floating occurs.

- Correct flare height
  For narrow-bodied aircraft this is indicated by a gentle sense of the runway rising to meet the aircraft just prior to touchdown. In strong gusty wind conditions, the appearance is more of flying onto the r/w.

...flaring at night!

Touching down off centreline

Off centreline, touchdowns are common after a pilot changes from right seat to left seat or vice versa. Changing from the RH to LH seat results in landings left of centreline. If this occurs ask the trainee to attempt to land on the side of the runway, he is tending not to use. Initial appearance or over-compensation will usually rectify the problem. Another help for a pilot flying from the LH seat is to manoeuvre the aircraft until it appears his right foot is on the centreline (vice versa for the RH seat).

Braking

Care needs to be exercised during manual braking. Ensure that the trainee's foot position is correctly placed on the pedals so that both rudder and brake can be operated simultaneously, especially critical during crosswind landings. Another potential problem is the type of footwear worn by the trainee. This author remembers well the problems that one particular trainee had during base training, with braking and keeping straight. A combination of small feet and slippery-soled shoes were to blame! Once the trainee had put on 'sensible' shoes, there was no further problem.

Pilot reports after landing

New pilots need encouragement to comply with regulatory requirements to report wind shear and wet runway braking action to the appropriate authorities. As a training pilot, you should research areas of knowledge and operations, which may be a little obscure or uncertain. If your trainee asks you a question or exhibits a problem for which you do not have an answer, admit it. Seek out advice and pass the answer on. Keep a library of articles and information that will help you in training. Study accident and incident reports to understand how they occurred so that you can pass on the experience to others.

The next chapter discusses the role of the base trainer and associated problems.

Flightdeck layout of Boeing 747-400.

9 Base training

The base training captain will have been carefully selected, vetted and checked, not only by the company that employs him, but also by the regulatory authorities, on whose behalf he will be authorized to undertake the legal duties assigned to him. It is of great importance for training captains to be first class aviators, and yet the best pilots are not necessarily the best instructors. The key to success in training is a high degree of professionalism in the special forms of communication, which help people learn effectively from the technical and human skill of the training captain. Learning, in turn, depends on the effective communication of skills, knowledge and attitudes. Regardless of the simplicity or complexity, the act of *communicating* is the vital key to the whole process.

However, the human being is so over-familiar with every day communication, that quite often he disregards the complexities of getting his point across effectively. There is a tendency to be imprecise, to generalize and let thoughts be poorly expressed. The effective trainer has to distance himself from this and use communication as a calculated and powerful tool.

During any period of instruction, trainer and trainee are involved in a constant interaction. The trainer giving the exposition on the appropriate topic, the trainee strives to assimilate the information. If assimilation occurs, learning takes place, if not then the trainee suffers anger, resentment and frustration, impeding any further learning. As a trainer you should ask:

- What learning the trainee needs to achieve
- What the trainer must provide to meet the trainee's needs.

The base training captain

Although most airline operators now carry out a significant amount of training in flight simulators, there is a continuing need for *in-flight* training. Modern flight simulators are now considered sufficiently realistic for all the items for an initial type rating test to be completed on an approved simulator. It is not uncommon for *experienced* pilots, having completed their simulator course, to fly their *first* real flight with revenue passengers.

However most regulatory bodies require certain items of the initial type-rating test be demonstrated on a specific type of aeroplane. Over the years, much criticism has been levelled as to the desirability of, for example, practising engine failures during base training. *UK Air Accidents Investigation Branch* (AAIB) points out that more accidents happen during such practice engine failures, after take-off, than real engine failures. The inference here is that if these *practice* engine failures ceased, then fewer accidents would happen. Of course the counter argument is that if pilots do not get the opportunity to practice on the real thing, then when a real engine failure occurs, they are badly equipped to deal with it, with even more disastrous consequences. Another report,[1] issued by the *French Commission of Inquiry*, investigating the crash of an Airbus Industrie A320, suggests that regulations governing airline pilot type conversion training, for the latest generation airliner flightdecks, are *totally inadequate*. The report states that during the certification of new generation aircraft there was no adaptation of existing regulations, or acknowledged methods, for improving standards, to take into account the ergonomic problems of the pilot-aircraft interface presented by the latest generation flightdecks. These accident reports draw attention of the inadequacies, not so much as to the regulatory authorities, but of the shortcomings, that can occur in airline flying training in general. The base training captain in particular needs to be aware of these potential problems.

In-flight procedures

During airborne base training, engine failure on take-off should *only* be simulated by reducing power and never by complete shutdown. When power failure is simulated during take-off, the speed should always be at or above V1 or take-off safety speed. The training captain must be alert, and ready to take over, if the trainee fails to maintain an adequate margin of control. Immediately before failure is simulated, the training captain should position his feet so he can prevent any application of wrong rudder input by the trainee. During and after simulation he must be particularly vigilant in monitoring airspeed, heading, pitch, roll, attitude and yaw. He must also carefully monitor engine instruments and aircraft configuration, especially on those types of aircraft in which a genuine failure of the idling engine would produce an abnormal hazard, for example long spool-up time.

---

[1] French Commission of Inquiry report into Air Inter crash Strasbourg 1992

Simulated engine failure on take-off can be a very marginal manoeuvre in crosswind conditions. The training captain must make sure that the speed chosen for the simulation allows an adequate margin of control. 15kts crosswind is the accepted maximum. A continued take-off following simulated engine failure during the ground run should only be practised in aircraft certificated in *Performance Group A*. On aircraft in other performance groups, simulation of an engine failure on the take-off roll must be followed by an abandoned take-off and should only be practised at a safe speed with sufficient runway remaining for the trainee to bring the aircraft to a halt.

Recommended techniques for simulating engine failure, of course, vary from aircraft to aircraft. Turbo-fan, turbo-jet, turbo-prop and piston engined aircraft will each have special problems requiring special techniques. For example, aircraft with turbo-fan engines, having slow spool-up times, it may be advisable to position the power lever slightly forward of *idle* in order to reduce response time if subsequent acceleration of the engine is required.

Actual engine shutdown

A pre-meditated shutdown, for most pilots, is a rare event. It is particularly important that refresher training is provided, preferably in a simulator, in order to practise appropriate drills and procedures. Emphasis must be placed on the need to conduct drills calmly, methodically and without haste. Whenever refresher training is carried out, involving an in-flight shut-down, training captains should be aware of the time required to restart an engine, in the event of an actual failure of a second engine during shutdown demonstration.

Recommended minimum safe heights for complete shutdown of power plants for training purposes

- Twin turbo-jet or turbo-fan engines   8,000ft
- Triple turbo-jet or turbo-fan engines   5,000ft
- Four engined aircraft   4,000ft
- Twin piston or turbo-prop   5,000ft
- Twin piston or turbo-prop (below 5700kg)   3,000ft

Preparation for flight

Due to the many psychological factors, a trainee pilot undergoing flight training is likely to be keyed up and tense, especially during the first few flights. As a tense person never reacts as well as a more relaxed one, the training captain should make every effort to detect tenseness and act to reduce this. Initially during flight training, trainee stress tolerance is likely to be low, but given proper guidance, assistance and encouragement this will improve rapidly as training progresses.

Briefings

Extensive briefings on the content of the flight session and its objectives are of paramount importance. The success of the training flight depends on the skill and communicative ability of the training captain. The training captain should provide a fully briefed lesson plan, following a logical step-by-step process. This will allay trainee apprehension and build confidence. A flexible plan will be required to accommodate the various factors, such as weather, airfield and air traffic control requirements. Proper briefing must precede the flight session ensuring that the trainee fully understands what he will be expected to achieve. Equally important is the post-flight briefing, in which all the executed manoeuvres are reviewed, faults analysed, praise given for well accomplished tasks and discussion on the areas that need further practice. As a trainer, it is often difficult to see and understand why trainees make mistakes. The trainee may see an error but take no corrective action, he may see it and take the wrong corrective action, or he may not consciously see an error and take no action. A good way to verify the trainee's perception and thoughts, is to encourage him to think aloud during a manoeuvre, thereby explaining what he sees and what action he is taking.

...think aloud!

Basic rules

- Power is used to maintain speed, *except* when thrust is a fixed value. The elevator is used to maintain altitude or to control the rate of descent
- In the fixed power setting case, as in take-off, climb, idle power descent, the airspeed is controlled by the elevator
- The primary means to counteract yaw during engine-out sequences is prompt but smooth application of rudder. Use ailerons for bank control only
- Trim should always, and only, be used to relieve control force pressure
- Under normal symmetric power and trim conditions, in most swept wing jet aircraft, use of the rudder pedals should not normally be necessary, except for certain aircraft types, where rudder is used for co-ordination of turns. The exception to this rule, for all types, is the crosswind landing.

> Under the first rule, *variable thrust*, it is recommended that the pilot have one hand on the power levers to react immediately to a speed change or to monitor the power lever movement when under autothrottle operation.

> Under the second rule, *fixed thrust*, the speed is controlled by the elevator, except of course when the aircraft reaches maximum pitch attitude.

> Under the third rule, *primary control for yaw is rudder*, rudder application to maintain direction must be prompt to maintain direction and offset the yawing motion. The natural tendency is to use aileron to control roll as a result of yaw. Rudder application prevents yaw and thus roll. Aileron must now be used to control bank.

> Under the fourth rule, *always be in trim*, pilots should be aware that trim is used to relieve pressure, not to initiate attitude changes. After a trim change, a short wait is required for its effect to take place. The aircraft should always be made to fly in trim.

Under the fifth rule, *little or no use of rudder*, is due to the yaw phenomenon that is inherent in swept wing aircraft. A pilot transitioning from a propeller aircraft, where possibly considerable amounts of rudder are required, will need to understand the strong roll characteristics associated with rudder usage.

Base training

Following an approved ground school technical training and simulator course, a trainee can typically expect to complete about two hours of airborne base training in preparation for licence issue. Once this base training is completed satisfactorily, the trainee is ready to commence his line training programme. Ideally, this base training should take place on two separate details to include day and night sessions. Flying training will cover those items listed on the flight training forms, issued by the airline training departments, to include all the mandatory items for the initial type rating. A good deal of these two hours will be spent carrying out visual circuits, with and without simulated engine failures, including touch and go landings.

Of necessity, base training could be said to be a fairly intensive, demanding, exacting and critical flying session, calling for exceptional skills in aviation resource management from the training captain. Many of the normal standard operating procedures will have to be adapted and adopted, to fit in with the training requirements. Extensive pre-flight briefings on these non-standard techniques are of paramount importance if the flight is to proceed safely. For example, touch and go circuits, not normally featured in day-to-day operations, demand a particularly high level of vigilance. The operational training value of touch and go circuits is somewhat dubious but they are usually necessary to avoid long, time-wasting taxi times, or hot brakes, due to repeated use. Ideally, two trainees would accompany the training captain, on a base training detail. One trainee would occupy the flight deck jump seat, he would be expected to operate the checklist on direction from the training captain, assist in monitoring the flight and generally act as a third pair of eyes. This relieves some of the load from the training captain, who then has more time to devote to monitoring both the trainee flying the aircraft, the aircraft itself, and maintaining liaison with ATC.

...extensive briefings!

Command authority

The training captain will of course have responsibility for the aircraft, however it is important the trainee has maximum command authority during the execution of specific manoeuvres. The trainee will be expected to give appropriate crew briefings, call for check lists, gear, flaps etc. and use the crew in an efficient and effective manner. If for any reason the training captain considers it necessary to take over control, he must clearly state *my controls,* which must be acknowledged by the trainee.

Crew co-ordination

The training captain should brief the whole crew if he considers a change in procedure is necessary. This possible change in procedure must be understood by all concerned to avoid misunderstandings.

Emergencies

If an in-flight emergency should occur, such as engine fire or failure, burst tyre, then the training captain would normally be expected to take over flying pilot duties, delegating second pilot tasks to the trainee(s). Prior to the first flight training session, trainees should be briefed on the procedures to be followed in case of a real emergency, including that of training captain incapacitation.

Incapacitation

The trainee pilot will normally have been trained on a flight simulator with visual aids, to an accepted level of proficiency and should be in a position to carry out the following:

- On approach - consider an overshoot - use autopilot
- Declare emergency - request assistance from ATC
- Decide on further course of action
- Do not hurry
- Consider re-seating second trainee pilot.

Aircraft characteristics

Whilst it is impractical to cover all possible aircraft characteristics here, there are some basic points to remember and emphasize during pre/post flight discussion. These relate mainly to large swept wing, multi-engined jet aircraft. Individual aircraft types will of course have their own peculiarities and will require additional briefing.

Low speed flying

Over two decades ago there were a number of fatal accidents to swept wing jet aircraft, it became apparent that these accidents could have been avoided, by revising the techniques used in the recovery at or near the ground under critical flight conditions. This resulted in most airline operators reviewing and revising their critical flight condition recovery techniques and methods of implementation for the so-called *second-generation* jet transports. The stall and recovery is perhaps, to the ordinary line pilot, of academic interest only, as it is not something he would encounter during ordinary line operations. However, as part of type conversion training all pilots are required to undergo stall recovery training, manoeuvres of which are never normally repeated during routine operations.

A stall and recovery is normally accompanied by loss of height. It is recognized that for *approach* to the stall, recoveries that result in rapid decrease in airspeed below bug setting, or rapid decrease in climb rate during take-off or go-around, or rapid sink rate on approach, then a more effective recovery method is necessary to avoid a potential disaster. For most aircraft under consideration the Vref (final approach speed) is usually 1.3 x Vs (stall speed) and on take-off, V2 (take-off safety speed) is usually 1.2 x Vs.

For such a critical flight condition, the *energy trade* principle is used. It is useful to know that if the Vref speed is around 130kts, and stall speed (Vs) is around 100kts, and stick shaker speed (Vss) is around 110kts then we have approximately 20kts of stored energy below bug setting before stick shake. This stored energy can and must be used if such a critical flight condition exists.

Factors affecting the stall

- Effect of momentum. Due to inertia heavy aircraft need more space and time to adjust to a new flight path
- Effect of power. Normally produces a pitch up, but due to inertia/momentum/spool up, it may take some considerable time before this is felt
- Effect of altitude. Indicated stall speed increases with altitude due to compressibility correction (difference between IAS and EAS) and actual EAS stall speed increases due to Mach number effect on wing
- Effect of weight. Stall speed increases with weight/G loading/bank angle.

Recovery from an approach to the stall

- In an emergency situation, positive climb performance and manoeuvre margins still exist at or near stick shaker
- Use aircraft climb capability below bug speed
- Smoothly control pitch attitude to minimize altitude loss
- Increase pitch attitude in small increments to prevent high pitch rates developing
- Intermittent stick shaker is the upper limit of useable pitch attitude to ensure manoeuvre and stall margins
- Pitch should not be increased so rapidly that airspeed decreases below stick shaker
- Apply maximum available thrust

- Maintain existing flap/slat configuration until positive climb performance then
- Accelerate to minimum manoeuvre speed for existing configuration.

...stall and recovery!

The latest, so-called *third generation*, of multi-engined jet transports mostly feature flight control computers (FCC) giving both minimum, *alpha floor*, and maximum speed protection, which effectively prevent the pilot from transgressing accidentally into the low/high speed regime. Flight control computers sense incursions into these areas and automatically take over, usually by autothrottle thrust increase or decrease.

Automatics have demonstrably enhanced flight safety, however this has not been without cost. All too often, the pilot is left outside the loop. *What's it doing now?* has been an all too common question heard on the flight deck. This is a reflection on the standard of many current training programmes, which leave the pilot poorly prepared in this type of environment. Training captains and trainees should be aware of these deficiencies. Know your aircraft and its capabilities, especially in the operating extremes of the flight envelope.

The base training programme

General points for emphasis

- Elevator trim. Short bursts of trim should be given to effect trim changes, as the stabilizer (most aircraft) is very powerful. Trim should be used to relieve pressure not initiate attitude changes (rule 4). Trim switches should not be used during take-off and landing phase, or to relieve back pressure during turns. Training captain to ensure that trainee is flying in trim otherwise workload increases with the result that other functions cannot be dealt with effectively
- Speed change. A change in speed may necessitate a relatively large change in attitude. For example a speed change of 10kts during final approach for aircraft like B747, MD11, DC-9 etc. require a 2° attitude change
- Sink rate. Low speed, high drag conditions on approach can result in high sink rates. Recovery in this situation can take a long time.
- Thrust setting. Be alert to speed trends and be prepared to make early and sufficient thrust changes on approach
- Speed/power stabilization. All approaches must be speed/power stabilized by at least a height of 200'. If not go around
- Flare. As sink rate increases, the kinetic energy of descent increases by the square of the sink rate velocity. To oppose this increased kinetic energy, lift must be increased considerably, otherwise a hard landing may result. Position of centre of gravity is an important guide on elevator sensitivity and effectiveness. Aft c. of g. elevator is lighter and more effective than forward c. of g. Note that training sessions are normally flown with an aft c. of g. When your trainee comes to fly line operations with a forward c. of g. the elevator feel at the flare will be appear to be heavier. This often results in hard landings on early line flying operations
- Touch down. Some low wing aircraft (MD-80 B717 etc.) produce a ground effect. This can have the result of not only reducing the sink rate at the flare by 200 - 300'/min, but also of pitching the aircraft nose down and if not taken into account may result in a long float. It is also important to appreciate that at the flare, just prior to touchdown, any increase in nose up pitch will merely drive the main undercarriage into the runway with a consequent hard landing.

Taxiing

Initially during taxiing there is a tendency for the trainee to focus at a point straight-ahead. This should be discouraged, especially whilst manoeuvring in close proximity to obstructions. Check wing and tail clearance, and make cautious thrust usage to avoid jet blast. It is a good idea to allow your trainee to taxy the aircraft in these early stages so he can get used to rudder pedal steering, and the use and effectiveness of the aircraft brakes.

Take-off and climb - common faults

- Tendency on take-off, for trainees sitting in the right hand seat, to position the aircraft to the right of the centreline, whilst those sitting in the left seat, position to the left of the centreline
- Runway centreline not accurately maintained with rudder
- Incorrect rotation rate
- In crosswind conditions tendency to use too much aileron in the early part of the take-off roll. With aircraft fitted with roll enhancing spoilers this can increase drag and induce yaw unnecessarily
- In crosswind conditions tendency at rotation to use insufficient aileron to maintain wings level
- Make sure aircraft is positively airborne and positive climb performance is obtained before commanding gear up
- Maintain correct pitch attitude, check aircraft in balance
- Anticipate acceleration/level change to avoid overshooting target altitude, especially on low level visual circuits.

One of the biggest problems confronting trainees is the speed at which things happen and change. They must be aware of the need to plan ahead, anticipate, and work to a carefully and fully briefed scenario. A thorough step by step briefing on what is required and when, will greatly smooth this early transition, reducing tension and allowing for a better performance.

Visual circuits

In normal line operations, visual circuits seldom have to be flown. Therefore, careful and well prepared conservative planning needs to be undertaken. Normally a take-off with subsequent circuit will only occur during flight training.

Visual approach

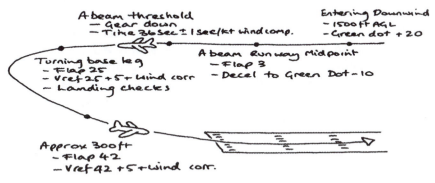

Abeam threshold
— Gear down
— Time 36 sec ± 1 sec/kt wind comp.

Entering Downwind
— 1500 ft AGL
— Green dot + 20

Turning base leg
— Flap 25
— Vref 25 + 5 + wind corr
— Landing checks

Abeam Runway Midpoint
— Flap 3
— Decel to Green Dot – 10

Approx 300 ft
— Flap 42
— Vref 42 + 5 + wind corr.

Typical low circuit - all engines operating

A constant look out is a very important when flying a circuit. Usually pilots spend too much time on instruments, not looking out frequently enough, and often realizing too late information that they would have recognized by outside visual clues.

Common faults

- Flying too fast in circuit
- Failure to allow for cross wind when planning circuit
- Insufficient bank used on base leg and final
- Overshooting centreline on final
- Incorrect descent rates ideally should be planned to stabilize at around 700'/ min. Ensure trainee scans vertical speed indicator frequently
- Aircraft not in trim
- Continuing approach when not fully stabilized
- Too much time spent head down on instruments with insufficient visual look out, resulting in poor positioning
- Too much or too little thrust giving unstable approach.

To be able to continue the approach visually the aircraft must be in a position where a landing can be made successfully. The speed/thrust/rate of descent must be stable. Insist the trainee goes back to basic flying techniques, where, if he is not happy with the approach then he must go around. A good *approach* is the key to a good landing. It must be realized that, unlike say a light aircraft, with heavy jet transport aircraft, due to inertia, it is impossible for a training captain to suddenly take control to rescue a poorly executed landing. Insist that the trainee use all the available information for glide slope guidance. During subsequent circuits as more skill develops then the training captain can gradually reduce the amount of glide slope information, for example by switching off the ILS, asking air traffic control to turn off PAPIS etc. The aim of the training captain is to get the trainee to a standard where he can undertake safe visual approaches using raw visual information.

Visual cues

Several factors in the visual environment can influence judgement and cause visual illusions during approach and landing:

Runway	Impression
long	high
short	low
wide	low
narrow	high
slope up	high
slope down	low
low visibility	further away
good visibility	closer in

Terrain in approach area	
high sloping down	low
low sloping up	high

Attitude control

To achieve effectiveness from swept wing jet aircraft pilots must pay particular attention to attitude flying. More precisely the relationship of thrust to attitude and airspeed. Remember the basic rules that condition flight when thrust is constant; the elevator is used to maintain target airspeed. During conditions of maintaining a fixed flight path, when thrust is variable, the elevator is used to establish and maintain the flight path, and the thrust is used to control airspeed. Training captains need to encourage their trainee to take a target speed and hold it as accurately as possible, thus facilitating attitude control.

Effects of controls

Often a neglected part of base training, causing trainees extra problems in basic handling and attitude control. Obviously each individual aircraft type will have its own particular quirks, and it will be the task of the training captain to ensure that the trainee knows the effect of the supplementary flight controls, thrust, flaps, slats, speed brakes, spoilers, gear.

Flare and touch down

A good landing requires a nicely stabilized short final with quick and accurate correction of any trend. The threshold should be crossed at 50′ wheel height towards the aiming point. The height of the flare depends on specific aircraft type, wind conditions, the aircraft weight and centre of gravity. As already mentioned an aircraft with a forward centre of gravity, *stabilizer trim aft*, the aircraft is more stable, requiring more stick force on landing rotation. Opposite for aft centre of gravity.

On a stabilized approach with a headwind, there will be a higher attitude nose up, and higher power setting due to the lower ground speed. In this situation fly the aircraft closer to the ground, reduce power slowly to idle just before touchdown. With a tail wind, ground speed is higher resulting in a lower attitude, higher rate of descent and less power. Here it is necessary to start the landing sequence earlier, take the power off faster and avoid floating.

Common faults

- Failure to make mental note of c. of g. position
- Incorrect speed at flare

- Aircraft decelerating while pitch kept constant, resulting in higher sink rate and possible undershoot or hard landing
- Harsh, unnecessary pumping of control column
- Landing to left or right of centre line
- Insufficient flare and level off
- Not looking far enough ahead to obtain correct perspective
- Wings not level
- Incorrect use of rudder causing yaw and subsequent roll
- Use of stabilizer trim during flare
- Failure to *land* nosewheel
- Late selection of reverse thrust
- Incorrect use of brakes.

In visual conditions the aircraft trajectory should follow that of the ILS aiming point, except when threshold clearance becomes marginal (some wide bodies). The trajectory established serves as a safe entry into the space over the threshold. However once the threshold is passed, the aiming point loses its significance and it is important that trainees realise the importance of looking well down the runway to avoid aiming point fixation. Be aware that judgement may be affected by other phenomena. Up slope in the runway or approach area creates an illusion of being above glide path, down slope creates an illusion of being below glide path. In hazy conditions expect to appear higher than you are. Bright lights appear closer and dim lights appear further away.

Touch and go

Touch and go landings demand a high degree of skill and co-ordination and must be properly briefed. By its very nature, a touch and go is a non-standard procedure never carried out on normal line operations. Its training value is therefore questionable, and has to be put in context with the necessity for this type of operation during initial base training. Namely avoiding time wasting, when full stop landings, accompanied by long taxi times and consequent hot brakes, may delay a subsequent departure. It is better value training, where if possible, touch and go landings are reduced to a minimum, so that the trainee can get the benefit of correct use of reverse thrust and brakes. This can often be achieved by adapting to local weather and airfield conditions. For example, in light or crosswind conditions, traffic permitting, it may be possible to make full stop landings, roll to the end of the landing runway, and take-off again in the opposite direction. Remembering if necessary to leave the gear down for a longer period to assist brake cooling. It must be clearly understood by the trainee, which type of landing is going to be made, and the entire crew must be correctly briefed on the procedure. There are basic guidelines to be followed prior to undertaking a touch and go, as without proper co-ordination between the crew, it can be a hazardous manoeuvre. Individual aircraft types may well have special requirements such as resetting flight director systems, take-off and go around switches (TOGA), flight control computers (FCC) etc. It is important that the training captain delegates the numerous tasks during this busy phase, after the aircraft has landed and prior to becoming airborne. These tasks will include briefing the trainee that his function will be to *fly* the aircraft, while the training captain will attend to peripheral tasks such as resetting flaps, trim, monitoring engine spool-up. Attention is drawn to the following points:

- Never attempt a touch and go after a simulated asymmetric approach, as there is no guarantee that the simulated failed engine will respond
- Do not arm ground spoilers, lift dump devices, or auto brakes unless specifically recommended in the operating manual
- No reverse thrust after touchdown. As reverse is *normally* selected, training captains need to be particularly vigilant to prevent this happening. If reverse thrust is inadvertently selected then consider making a full stop landing, as there may be insufficient runway length left
- Immediately after landing, be aware the aircraft will not slow in the normal way and will appear to be very lively. The trainee must keep his undivided attention to keeping on the runway centreline, with the wings level, whilst the training captain will be attending to resetting trim, flaps and engine spool-up.
- It is possible and quite likely that whilst the flaps, trim and engine are being adjusted, the take off configuration warning will sound. It is important that the trainee is made aware of this and is not distracted from his primary task
- The speed bug setting for the approach may differ from that required for take-off
- Practice flapless approach and landings should normally be followed by touch and go, to avoid high stress on tyres and brakes

- Ensure that you know your individual aircraft operating techniques regarding the setting of FD, FCC, and TOGA.
- It must be fully understood, by all crew members, who does the actions in the case of an actual failure during a touch and go.

Go around

The go around is an essential manoeuvre, to practice and emphasize, during initial base training. The trainee will have had a good deal of practice doing this in the simulator, but mostly under instrument flight conditions. This is the opportunity to *hone up* his skills and build confidence, which may one day prove invaluable in normal line operations. Several practice visual go arounds, from different heights and configurations, should be attempted, until the trainee is competent at this manoeuvre.

Common faults

- Under rotation. The trainee should realize that the only way to stop descent and climb away is the correct pitch attitude
- Incorrect thrust setting. Be particularly aware for overboosting
- Wrong commands
- Forgetting gear/flaps
- Incorrect use of trim
- Distracted from primary task of flying aircraft
- Failure to observe correct go around speed
- Failure to keep aircraft in balance, especially in the case of engine out go around.

Wave off

This is similar to a normal go around except that the manoeuvre begins at the flare with the thrust levers at or near idle. The normal procedure is to apply go around thrust, rotate to arrest the sink rate, make no changes to aircraft configuration until positive climb performance has been achieved. Be aware that the main gear may touch the ground during this manoeuvre.

Crosswind take-off

As already mentioned, a crosswind can cause some difficulties during early conversion training. It is worth emphasizing to the trainee the technique to be adopted. Remember swept wing, high tail aircraft have a tendency to steer into wind on the ground roll and heel over at rotation. Slightly more forward pressure will assist in maintaining directional control and some wheel input will be required, as the upwind wing will have a tendency to rise. After lift off the desired track should be maintained by crabbing.

Common faults

- Too much wheel throw during early part of take-off
- Insufficient wheel throw on lift off to keep wings level
- Too much forward pressure on control column, causing *wheelbarrow* effect
- Not maintaining correct track during climb out.

Crosswind landing

There are two methods of crosswind landing techniques. Both have their advantages and disadvantages. The method to use will vary, depending upon runway braking action state and aircraft type. See the aircraft operating manual for specific guidance on which, or both methods if applicable, to use.

Sideslip crosswind technique

The objective of the sideslip technique is to hold the longitudinal axis of the aircraft with the centreline, during the final phase of the approach and touchdown. At approximately 50′, progressive downwind rudder should be applied to align the aircraft's axis with the centreline. As the rudder is applied the upwind wing will sweep forward, causing roll, this must be corrected by application of aileron. Use the rudder to hold the aircraft's axis parallel and the ailerons used to keep wings level. It is important not to commence rudder input too early, otherwise by the time the aircraft reaches the touchdown point it may well be subject to further drift away from the centreline. Equally, it is important that the aircraft is not landed with drift on, as side loads on the undercarriage are undesirable. Bear in mind that some low wing aircraft will touch wing tip or engine pod first at bank angles in excess of 8°.

Partial crab crosswind technique

This is often the optimum method favoured, especially on runways that are wet or contaminated, or where the tyre adhesion and braking action may be less than good. The technique is to land with part crab and part wing down. At the initial landing roll, on a slippery runway, almost no side adhesion of the tyres is available, until the speed is below aquaplaning speed. With no pilot corrections, it is likely the aircraft will tend to weathercock into wind, and drift towards the downwind side of the runway. Any application of reverse thrust in this situation will only aggravate this drift. It is important if directional control becomes a problem, then reverse thrust must be cancelled and sufficient forward thrust maintained until wheel spin up and tyre adhesion takes place. Implications of the increased stopping distance need to be considered.

Common faults

- Aircraft not on centreline before, during and after de-crab
- Insufficient allowance for long bodied aircraft. Main wheels to follow runway centreline
- Rudder and aileron inputs incorrectly co-ordinated
- Late aileron input
- Excessive bank
- Touchdown on down wind wheels first
- Overcontrolling.

A useful exercise can be demonstrated, during crosswind conditions, by allowing the trainee to fly level, at about 100', along the entire length of the runway, maintaining the centreline and runway heading. Here the trainee can get first hand experience of the required rudder and aileron inputs and see the importance of maintaining steady control inputs.

Braking procedure

The base training captain needs to pay particular attention to briefing his trainee on the correct use of brakes. Careful attention here will pay dividends and avoid the possibility of burst tyres, or overheated brakes. Prior to the first take-off, the trainee should be allowed to try the brakes, during the taxiing phase, so it is not a surprise after the first landing. Attention to feet position, and even the correct type of footwear, are important and are often overlooked.

Common faults

- Too fast on approach, resulting in long float and brakes applied too early
- Application of brakes before wheel spin up resulting in locked wheel
- Too smooth touch down, resulting in late wheel spin up
- Asymmetric braking
- Harsh brake application
- Brakes not applied smoothly and consistently
- Failure to continue braking throughout the deceleration phase
- Insufficient force applied on brakes.

Engine failure manoeuvres

Before engines out manoeuvres are performed in the circuit, the trainee must be proficient and confident with all aspects of the normal, all engine-operating procedures. Normally an engine failure is simulated by smoothly retarding the thrust lever. Actual shutdown of an engine should not be considered, except for the purpose of training the shut down and re-light procedure when this would be done at a suitable altitude *(see table p182)*. It should be pointed out to the trainee, that as the thrust lever is moved smoothly to idle, the flying characteristics will be different to that of a sudden engine failure or seizure. In addition, where this manoeuvre is carried out immediately after V1, the trainee should be aware that the training captain would maintain his hands on the thrust levers. A full briefing on these manoeuvres will prevent any confusion. As a safety measure, it is recommended that with engines having long spool up times, then the simulated failed engine should be partially spooled up ready for use should the other engine fail.

Trainee reaction to engine failure should trigger the same sequence of thought and reaction. Application of rudder to counteract yaw, control of the flight path, airspeed and pitch, with a subsequent call for follow up action by pilot not flying will be required. The task of the training captain will be to make this exercise as realistic as possible, insisting on standard operating procedures and flight profiles.

Theory of flight with asymmetric thrust

If during asymmetric thrust the aircraft is flown wings level and ball centred, it is in a steady slide slip, and rudder effectiveness is reduced. Improved rudder effectiveness, lower drag and lower Vmca, can be achieved by banking the aircraft a few degrees towards the live engine. Here the ball will be slightly off centre, towards the live engine, and flight will be at zero slip with less drag.

Aircraft with wing mounted engines require more bank than those with aft mounted engines. However too much bank increases Vmca due to flow separation at the vertical fin. At high weights, performance requirements may be critical, the V2 speed will be well above Vmca, and consequently the bank required to fly at zero slip is small. However at low weights V2 is nearer the Vmca and is therefore more critical from a directional control point of view. Flown wings level, performance penalty is greater, but due to the lower weight, the resulting performance should not be critical.

Asymmetric flight with wings level, simplifies instrument flying procedures, which will give the average pilot higher aircraft performance than the more sophisticated procedure using bank. However, at the common low weights used in base training, care must be taken to maintain directional control.

Summary

Apply rudder and fly wings level. When rudder input is correct very little aileron is necessary. If directional control is a problem then bank a few degrees towards the live engine. With four-engined aircraft, a double engine failure on the same side will present directional control problems and bank towards the live engine(s) will normally be required. Circuits with double engine failure should only normally be practised in the simulator.

Note that by definition Vmca during certification includes 5° bank.

See Vmca v angle of bank diagram overleaf.

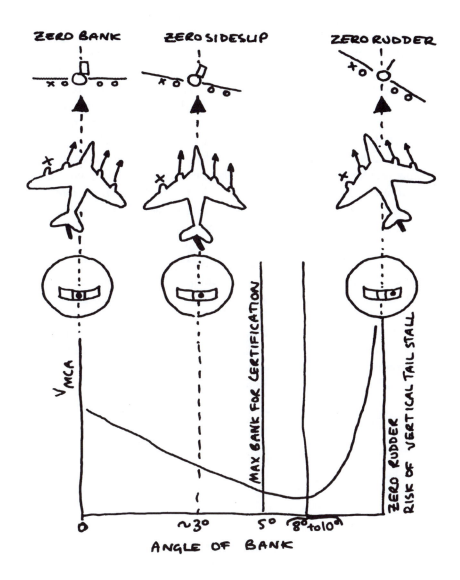

Vmca v angle of bank

Engine failure during take-off

Normally the trainee will have practised numerous engine failure drills and procedures, during the simulator conversion training. It is a fundamental requirement that the pilot is fully able and competent to cope with an engine failure just after V1, maintain directional control, climb out at V2, accelerate to en-route climb configuration, action the appropriate drills and procedures and carry out an engine out circuit and landing or go around. As mentioned earlier, many training accidents and incidents have occurred during base training operations, which might have been avoided by more careful planning, briefing and application. The simulated engine failure on take off requires a good deal of skill, co-operation and co-ordination from both the training captain and the trainee, if it is to be conducted safely.

As a safeguard, it is recommended that the training captain should place his feet lightly on the rudder pedals, not only to confirm correct rudder application but also to prevent wrong rudder inputs, which on some aircraft could have disastrous consequences. In crosswind conditions extra care needs to be exercised, both in choice of simulated engine failure and the speed and point at which the simulation takes place. From the directional control standpoint, the upwind engine is usually the most critical, as both the asymmetric thrust and the crosswind component, are forces acting in the same direction. On the other hand, an engine failure on the downwind side may necessitate, either no rudder at all, or opposite rudder in order to keep straight whilst still on the ground. An additional complication, that may require a two stage rudder input, is an engine failure where the nose wheel is still in contact with the runway. At rotation, the directional force exerted by the nosewheel will be reduced, requiring additional rudder input.

Common faults

- Trainee anticipates failure and consequently acts prematurely
- Failure to keep straight on runway centreline
- Not enough or too much rudder application requiring additional aileron input
- Incorrect rotation rate
- Constant pressure not maintained, causing roll due to yaw
- During clean up, lets aircraft sink
- Adopting wrong profile for climb out.

Engine out approach

This should only be attempted after the trainee has demonstrated his competence at making normal approaches. Briefings should include the differences of flap settings, profiles, pitch attitudes, power settings, rudder control and flare and landing technique. Rudder trim should normally be zeroed before landing. Additionally the training captain should ensure the simulated failed engine is spooled up, ready for use should a real emergency occur.

Common faults

- Unstabilized approach
- Late rudder application. Rudder should slightly lead thrust application to prevent dynamic yaw
- Insufficient power changes to compensate for engine out
- Failure to appreciate different nose attitude on approach due reduced flap setting
- Forgetting to zero rudder trim before landing
- Failure to maintain runway centreline.

Engine out go-around

Common faults

- Insufficient rotation
- Insufficient or late power application
- Go around/*TOGA* switches not selected
- Poor look out
- Poor instrument scan
- Failure to follow prescribed go around procedure
- Overboosting engines
- Aircraft not in balance
- Failure to comply with *ATC* go around procedure
- Calling for gear/flaps before positive rate achieved
- Failure to call gear/flaps up.

Type rating requirements

As part of the airline operator's certificate (AOC) each regulatory authority, will normally approve the company's training schedule, provided it meets their minimum mandatory requirements. The complete conversion training syllabus would normally be contained in the company's operations training manual. As each item of the conversion-training course is completed, the appropriate trainer should sign it off.

Typical requirements for pilots not qualified on type would be as follows:

- Approved ground technical course, followed by examination
- Cockpit procedure training, operating as a crew, typically 10 hours shared
- Flight simulator training, on an approved simulator, typically 32 hours shared
- Emergency and survival training
- Performance course
- Flight planning course
- Loading, trim and loadsheets
- Base check and instrument rating renewal
- Pilot incapacitation
- Initial line check
- Airborne base training
- Line training
- Final line check.

The airborne base training element will typically consist of approximately 2 hours, except where experienced pilots have completed Level 4 simulator training, on a suitably certificated simulator, in which case it is normal for the trainee's first flight to be on line revenue operations. All items on the application form, for the inclusion of the specific type rating, will be required to be completed satisfactorily and signed off by an authorized trainer. Most regulatory authorities require that all the items in the above list, up to the completion of base training, are completed satisfactorily, prior to type rating issue on the trainee's licence, before line training can commence.

The proficiency check and licence renewal

The company proficiency check and instrument rating is normally conducted as part of the bi-annual check and training session in the simulator. Although some operators are now advancing their philosophy in advanced qualification programmes. *(see Chapter 14, Total qualification programme).*

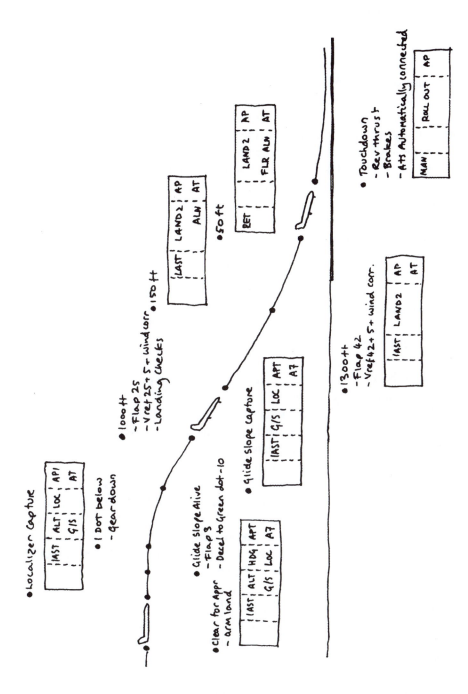

Typical engine failure approach

The line proficiency check

Each pilot is normally required to undertake an annual check flight during normal routine line operations. Check pilots will need to exercise a high level of common sense and discretion when assessing points which may arise, bearing in mind that the main objective is to attain and maintain, the highest possible standard of everyday operations. The check flight gives a good opportunity to see and assess the crew in action, under real airline conditions. All aspects of the operation should be considered, including aircraft systems and knowledge, aviation resource management, standard operating procedures, abnormal and emergency procedures, emergency equipment and overall crew co-operation and communication. The check pilot should ideally adopt the role of facilitator, offering support and encouragement where appropriate, and avoid being seen as a *trapper* or company *snoop*.

Pilot assessments

What qualities make a good airline pilot?

He should be stable, reliable, be able to stand up to pressure, have aptitude, flexibility, an ability to get on with people, be well motivated, have perseverance, energy, wide interests, initiative, independence, emotional maturity, and be capable of working under high stress levels. Above all perhaps, flying should be considered as controlled disciplined fun, bringing with it a new dimension of creative, infectious enthusiasm. How can we assess these qualities? In broad terms, we need to know: How able is this person? What *kind* of person is he? Answers to the first part can relatively easily be measured, by performance related tasks, aptitude and intelligence testing, and with existing pilots during routine base/line training. Answers to the second part are less easily defined objectively. When assessing personality it is important to study the so-called *softer* human factor issues. Namely the interaction with others, thinking style, feelings and emotions, empathy and imagination, an ability to think clearly, act decisively, accept responsibility, use initiative, work effectively and demonstrate the right temperament, and to control the aircraft and crew under normal and emergency conditions. Crew co-ordination, appearance, bearing, judgement, confidence, competence, and capacity, present the training pilot with one of the biggest challenges in assessing these qualities.

Progress reports

It is an inherent part of the training pilot's task to compile progress reports and objective profiles on his trainees. Certainly no easy job! Look now at the three pilot progress reports created below: only one, *the latter*, has any useful place in objective, positive reporting.

Pilot Progress Report
Name......*B Atwerp*...............**Rank**......*Capt*......
Fleet.......*B747/400*.............**Licence**......*AT/316234/A*
Proficiency/Line/Special Check*   Pass/Partial/Fail*   **Delete as appropriate*
Date......*01/06/01*......................
Remarks/Recommendations   *A very good performance.*   Signed.......*ABerke*   Name (Print)....*A. BERKE* ..(Trng Capt.)
On completion this form will be addressed to the relevant Flight Training Manager under "Confidential" cover and filed with the pilot's training records. A copy will be forwarded to the candidate.

Pilot Progress Report

Name......*B Desparate*...............**Rank**......*F/O*......

Fleet.......*B737*..............**Licence**......*AT/675431/T*

Proficiency/Line/Special Check*   Pass/Partial/Fail*

*Delete as appropriate

Date......*01/06/01*.....................

Remarks/Recommendations

This is the worst Proficiency Check that I have ever seen!
F/O Desparate gave a poor take-off brief and flew the departure to an unacceptable standard. The initial circuit to the ILS was not much better.

A further ILS was all over the place culminating in a go-around which I thought was dangerous.
He had great difficulty with the non-precision approach and failed the holding pattern.
He gives the other crew members no-confidence.

A repeat of all items is required.

Signed.......*ABerke*   **Name (Print)**....*A. BERKE* ..(Trng Capt.)

On completion this form will be addressed to the relevant Flight Training Manager under "Confidential" cover and filed with the pilot's training records. A copy will be forwarded to the candidate.

Pilot Progress Report

Name......*Snodgrass*...............**Rank**......*F/O*......

Fleet.......*A320*.............**Licence**......*AT/6578490/G*

Proficiency/Line/Special Check*   **Pass/Partial/Fail***
                                        *Delete as appropriate

Date......*01/06/01*......................

Remarks/Recommendations
F/O Snodgrass did not achieve the required standard in the following areas:
1. General aircraft handling problems caused by lack of accurate rudder trimming, compounded by a slow instrument scan rate. These factors resulted in many significant excursions from the desired horizontal and vertical flightpath, together with inaccurate speed control.
2. Lack of assertiveness, coupled with low volume of speech portrays a lack of self confidence,
I recommend a period of general handling practice concentrating on accurate trimming and increased instrument scan.
Careful revision of SOP's and briefings are necessary as a foundation for correcting many errors.

Continues to maintain a positive attitude despite these shortcomings. Appearance immaculate as usual.

Signed.BGoodwun **Name (Print)** GOODWUN (Trng Capt.)

On completion this form will be addressed to the relevant Flight Training Manager under "Confidential" cover and filed with the pilot's training records. A copy will be forwarded to the candidate.

It is the training pilot's function to *train*, offer help and inspiration to his trainee, to set targets, and give positive feedback. It is useful to have a standardized marking system. This not only helps the objectivity of such reports but importantly gives the trainee positive guidance on his perceived strengths and weaknesses. *(see Chapter 14, Total qualification programme).*

...total qualification programme!

Base check and instrument rating

In most companies, this is carried out in a full flight simulator. In the UK, this now forms part of the bi-annual base check training requirement and annual instrument rating renewal.

Conclusion

The base training captain has the delicate task of both training and testing. If he is to be effective then the emphasis should be *Training* with a capital *T* and *testing* with a small *t*. As we have already seen there are many aspects of flying a modern commercial aircraft that need special skills in identifying problem areas and imparting this information in a helpful and positive way, both to the trainee and the airline training department. The next chapter discusses the techniques and problems associated with command training.

During any period of instruction, trainer and trainee are involved in a constant interaction.

10 Initial command training

The depth of training required for new captains will be considerably deeper than a normal upgrading situation. Most importantly an initial command trainee must be given and exposed to the responsibilities of command. There are many aspects of commanding a modern commercial aircraft that are not always readily apparent, even by long-experienced first officers. In fact, many of these first officers fail their command assessments due to the fact that they have been in the right hand seat support role for too long and have become unaccustomed to the complexities of command and decision making.

When an initial command trainee is required to undergo command assessment, he or she should be provided with as much information as possible for successful completion of the required task. Aircraft captaincy involves the management of the aircraft, its crew, and the supporting company infrastructure, in an ever-changing environment, if it is to achieve the highest levels of safety punctuality, comfort and efficiency. The initial command trainee is at a high point in his career. His motivation and enthusiasm will be at a peak.

The trainer may find that his major task is at first to direct his trainee's energies to constructive effort instead of, for example, hours spent compiling mini manuals, note books, and beautifully coloured SID and let-down charts. The use of aids such as these should not be discouraged, merely kept in perspective. Indeed, the trainee is often confused by the wider scope of knowledge he now perceives as necessary, and his priorities may become disordered. This situation is by no means the norm, however the trainer should be alert for it, as it can amount to wasted effort and be an early indication that the trainee is having difficulty in ordering priorities.

From the first day, the First Officer, in his new role as a Captain under training should accept that he is the Captain. The trainer can assist here, by referring all decisions to him, not influencing his decision unless absolutely necessary, introducing him as the Captain to flight attendants and ground personnel and by placing his name first, in the flight report. During the early stages of training, the trainer should be as supportive as required, working always to lessen this support until it has been withdrawn completely.

Throughout the history of most airlines, it has been very rare for a pilot not to clear to line because of deficiencies in flying ability. Where problems have manifested themselves, it has been within the domain of flight management. This should be the subject of attention throughout the training programme, and the trainee should have a clear understanding of his priorities throughout the flight. Company policy of *safety, schedule, comfort, and economy* should be reinforced. Further, action following an emergency or abnormal situation, through to landing or evacuation of the aircraft should be known, along with an appreciation of performance, range or endurance following one set of occurrences. Restate the policy of, *fly the aeroplane first,* whilst having regard to the concept that total performance of the crew is greater than the sum of the performance of the crewmembers taken independently.

...fly the aeroplane first!

Being in command of the situation at all times with sound integration of human effort and available resources is the aim. To lose this thread through cockpit intrusion, disruption, or a breakdown because of overload is something the trainee should be constantly on guard against. Soundly applied *flight management* principles should be number one priority at all times.

Instructional climate

An important task for the trainer is to maintain the instructional climate within the range where learning can take place. Keep the levels of stress within bounds.

Dealing with problems of crew co-ordination is rather a different matter from normal technical instruction. There are no straightforward precise limits or indicators in the abstract field of leadership and management. This so-called *grey area* is complex and somewhat indefinable, as it is measuring subjective observations, based on values, impressions and feelings. This can present the trainer with difficulties in making accurate assessments and judgements, especially so with initial command assessment training.

The training pilot must never assume that the perception he has of a given situation is necessarily shared by the trainee captain. He should diagnose, seek information and ask open questions, aimed at finding out how the trainee views the situation. This process should lead into a problem-solving mode of seeking more information, of summarizing, of making positive evaluation and giving positive feedback. In any given constant conditions, individuals will vary from moment to moment in learning rate.

Plateaux and even regression can arise naturally through mental overloads. Poor ineffective instruction too, has the ability to create a plateaux or regression. Rates of learning will be influenced by fatigue and normal diurnal rhythm. It is well to remember that almost as soon as instruction stops the forgetting process begins.

What have I done, What haven't I done, What needs to be done, should become cockpit oracle of the Captain.

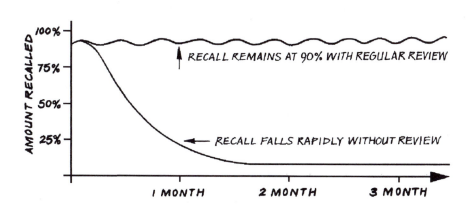

Rates of learning

Development

The initial command trainee will be well motivated and keen to produce positive results. He may well be apprehensive as to what is required of him.

The trainer can help enormously here by:

- Take logical steps, have a training plan, covering the required syllabus *(see pre-command training and assessment at the end of this chapter)*. Remember what the trainer has to say must be logical to the trainee, whose logic may be different from the trainers. Assume nothing and move in simple steps
- Create memorable steps, impressions should be proportional to the importance of the points the trainer wishes to make
- Check progress, ensure the trainee is made aware of any shortcomings at all stages of training
- Use knowledge and experience make use of the trainee's expertise, to anchor new material
- Confirm results, essential that the trainee gets positive indication how he is progressing and has a clear picture of his overall progress
- Give feedback to the trainee so that he understands, is able to accept the information and do something about it.

The ability to recall information decays rapidly if the information is not used memorably and in context. However, repetition can go too far causing irritation and memory blocks. A concise well-placed summary can help fix the general shape of the subject in the trainee's mind. Repeated recall and use of information in different ways can boost memory significantly.

...achieving high performance!

There is a great deal of evidence that high levels of performance are influenced and developed by an ability to *image* the performance in great mental detail, carried out correctly and in real time. Practice, rehearsal and asking *what if?* will expand both the trainee and the trainer's abilities in a conscious and deliberate way. Acute powers of observation, the facility to relax and concentrate on the whole performance, with a positive attitude are the keys to success in this area.

- Produce intense expression of activities through word pictures, illustrations, demonstrations and wide use of senses
- Allow and encourage the trainee to practice and rehearse so that he can develop his capabilities in a deliberate and conscious way
- Ensure that a positive self image is fostered whilst avoiding overconfidence
- Ensure that such practice is correct in every detail but not at the expense of flexibility.

Imaging can only be developed through a two way communication process. Every opportunity should be taken whilst giving instruction, direction or explanation, in such a way, whilst economising on time, to evoke strong imagery and clear logic. Illustrate, demonstrate, practice and give positive evaluation. As already mentioned *(see Chapter 4)* effective communication must always contain non-verbal, as well as verbal behaviour. Body posture, hand gestures, facial expression, head nods, eye contact, all play an important part.

Feedback

This should be an immediate aid to the trainee in finding new or more effective responses.

- Perceptions, reactions and opinions should be presented as such, and not as absolute facts
- It should refer to the relevant performance behaviour or outcomes and not to the individual as a person
- It should be in terms of specific, observable behaviour and not in general terms
- When used in evaluation rather than descriptively, it should be in terms of established criteria, probable outcomes or possible improvement, as opposed to making judgements such as good or bad

- Regarding performance it should include discussion of what is viewed as the high and low point of that performance
- Problem areas in which there are established procedures for achieving solutions, suggestions should be made regarding possible means of improving performance
- Avoid loaded terms which produce emotional reactions and hostile defences
- It should be concerned primarily with those things that the individual has some control, and be given in such a way so that it can be used for improvement or planning alternative actions.

...avoid feedback giving emotional reactions!

Pre-command assessment training

Prior to command assessment, it is recommended that the initial trainee captain complete a comprehensive classroom training course, covering all the aspects of command management and company organizational culture followed by a syllabus of supervised line flying, with an experienced training captain. A typical syllabus will include twenty sectors followed by a command assessment period of around six sectors. During the twenty sector preparation period, the nominees will be encouraged to handle as much of the command decision making process as possible, in order to equip themselves with the necessary skills and knowledge prior to command assessment. The logical implementation of the sequence of learning will largely be predicted by events unfolding on the actual sectors flown, but every effort should be made to ensure that the syllabus is fully covered within the allocated number of flights. It is therefore hoped that individuals will be better equipped for command assessment, and that the knowledge gained will provide a secure foundation for a smooth progression to command.

The following pages[1] of this chapter indicate the items that should be covered during pre-command assessment training for a typical short haul airline operation.

Weather brief interpretation

- Areas of clear air turbulence - cabin service/staff liaison

- Thunderstorm Cb activity - cabin service/staff liaison

- Weather radar – availability

- Crosswinds - runway availability, diversion availability, fuel requirements, runway states, approach aids

- Runway performance, contamination, length, take off speeds, landing distance, take off and landing technique, use of flaps, use of APU

- Low visibility - take off and landing requirements, extra fuel, aircraft serviceability, Cat 1/2/3, approach aids, diversion airfields

- Take off alternate

- En-route icing - extra fuel requirements.

Aeronautical information bulletins

- Take off, take off alternates, en-route alternates, destination and alternate aerodrome opening hours and approach aids

- ❏ Required navigation aids en-route and landing available

- ❏ New departure and approach procedures

- ❏ Fire category/cover available

- ❏ Runway length available

- ❏ Work in progress.

Cabin staff pre-flight liaison

- ❏ Introduction of all the crew

- ❏ Flight conditions

- ❏ Flight times

- ❏ Slot times

- ❏ Delays.

Ground staff liaison

- ❏ Ordering fuel

- ❏ Technical problems

- ❏ Passenger boarding times in conjunction with c/s

- ❏ Transport to aircraft

- ❏ Loadsheet requirements, extra cabin/flight deck, airstairs, spares carried, freight loading

- ❏ Dangerous goods

- ❏ De-icing requirement

❑   ATC start delays.

Engineering liaison

❑   Technical log availability

❑   De-icing complete and inspected correctly with correct tech. log entry

❑   De-icing protection, hold over times

❑   Tech. log examined - CFD's, notice to crews, special inspections, C of M review

❑   Defects arising during pre-flight checks - rectification possibility and time, correct entry in tech. Log.

Cabin staff liaison on the aircraft

❑   Technical delays

❑   ATC delays

❑   Unexpected turbulence - seat belt sign, PA

❑   10 minute landing warning

❑   Extra time available for service due holding

❑   Unexpected decrease in flight time due re-routing

❑   Disruptive passengers - action before, after and during flight

❑   Emergencies - chain of command, expected landing time, cabin staff required action, update of situation, information from cabin staff

❑   Evacuation procedure

- Cabin staff under training

- Meal breaks - which turnaround? how long?

- Security checks

- Boarding time

- Cabin defects

- Refuelling with passengers on board

- Diversions

- Expectant mothers, special facilities for disabled pax

- Sickness aboard - medical facilities.

Passenger liaison

- Boarding PA

- Delay PA

- Refuelling PA

- Turbulence PA

- En route PA

- Emergency PA

- Technical problems - pax to remain on board? drinks food service, trolleys in aisles, evacuation problems?

- Missing passengers, baggage rationalisation during technical delays. Staff liaison, transport to terminal

- Life jacket, safety equipment demonstration

❑   Jump seat passengers - briefing, min. age, alcohol.

Company liaison

❑   Use of company frequency

❑   Differing frequencies for normal messages and radio closeout

❑   Information required

❑   Handling agents - charter brief

❑   Sensitive information

❑   Security information - muffle code

❑   Hi-jack line of communication.

Crew duty hours

❑   Extension of duty - flight report

❑   Reduction of rest - flight report

❑   Differences for cabin staff

❑   Maximum duty period

❑   Split duty

❑   Meal times

❑   Flying duty period vs company duty period

❑   Surface transport times to hotac

❑   Hotac rest period

- Hotac on long turnarounds

- Reduced report times

- CAA reports required.

Carriage of goods

- Firearms - where? ammunition, how much? where?

- Diplomat/VIP protection

- Stowage of dangerous goods

- Human remains

- Transplant organs

- Engineering spares

- Pacemakers

- Radios, cassettes, mobile telephones, computers

- Company mail

- Animals - international flights.

Paperwork

- Tech. log - CFD transfer, downgrading to Cat1, signing off defects, de-icing, pages to be removed, lightning strikes, PDI, fuel discrepancy

- Customs crew declaration - compilation

- Dangerous goods - signing for

- Birdstrike forms

- GPWS forms

- Airmiss forms

- MOR forms

- Death on board – action.

Emergency equipment

- Minimum equipment to be carried

- Life jackets/babycots - requirement over water

- Oxygen requirements

- Fire extinguisher requirements

- Inoperative exits.

Diversions

- Preferred alternates

- Onward passenger transport

- Handling facilities - fuel, ground power, de-icing rig, airstarts, loadsheet, hotac, credit facilities

- Maintenance facilities

- Geographical position

- Political ramifications

- Crew replacement - duty hours.

Emergencies

- ❑ ATC

- ❑ Cabin staff liaison

- ❑ Passenger PA

- ❑ Company liaison.

Aircraft performance

- ❑ State minima

- ❑ Approach bans

- ❑ Circling minima

- ❑ Minimum safe altitude

- ❑ Terrain clearance in turbulence, driftdown

- ❑ Commanders absolute responsibility for terrain clearance

- ❑ First Officers handling limitations

- ❑ Airport/area clearance certificates

- ❑ Company operations manual requirements

- ❑ Fire cover.

Legislation

- ❑ Designated aircraft commander

- ❑ First Officer responsibilities

❏   Crew responsibilities.

Minimum equipment list

❏   Interpretation/implementation.

External check

❏   All accessory compartments shut, E&E bay hatches shut

❏   Spare wheels lashed down

❏   Positive visual and tactile information of upper wing surfaces in suspected icing conditions

❏   Gear pins removed.

Refuelling

❏   Normal refuelling including aux. tanks, wings first, pre-selection

❏   Manual refuelling

❏   Overwing refuelling

❏   Transferring asymmetric fuel load

❏   Carnet, validity

❏   Dripless sticks

❏   Water drains

❏   De-fuelling.

Starting

- ❑   Air truck assisted start

- ❑   Battery start

- ❑   Cross bleed start, precautions, effect of air. cond.

- ❑   Electrical X tie problems

- ❑   Hung start.

The next chapter is involved with pilot selection, career appraisal, development, and suitability.

---

[1]Pre-command Assessment Training Form. Courtesy of British Midland Flight Training Department.

Not yet the paperless aircraft.

11 Pilot selection

In general the great divide between the most effective and least effective instructors, appears to revolve around whether learning is treated as a one-way process, or a two-way process. As a two-way process, effective instructors exhibit a balance between activities of *feeding* and *demanding*. They are sensitive to when and how much they should *feed* their trainees and when they should *demand*. This can be achieved by setting challenging tasks, asking them to explain their actions, and obliging them to make decisions. There are many negative characteristics, common to the styles used by some instructors, which are evidently ineffective.

Inherent in every training situation there are requirements which the instructor must meet in his own way. That personally chosen style will exercise a wide influence, conditioning the way each trainee views the organization, the company, and the job itself. It will quite possibly shape the climate of the airline training department and how it is perceived by the workforce. Selecting good quality training pilots is of obvious importance if pitfalls are to be avoided. An awareness of what is possible and desirable, can in itself, bring about the first steps in contributing to effective training, and improvement to working life.

The need for flight instructors to be trained in instructional skills has often been repeated. There are very special circumstances in flight operations that have to be acknowledged and utilised if instruction is to be effective and efficient. A good general background as a qualified flying instructor, at both basic and advanced level, is considered the minimum starting point for anyone contemplating a career in airline pilot training. So many of the skills, applied during basic flying training, are highly relevant to the airline world. Indeed basic raw flying skills, learnt and applied at ab-initio level, are the mainstay of all flight operations. All too often in the airline training world this basic skill becomes obscured and forgotten.

The flight deck provides a unique training environment. Complex technical systems, operation to company procedures, a highly specialized and designed workplace, speed, accuracy, flexibility, performance, all of these and many others, make it a special training challenge, requiring dedicated, motivated and well trained professionals.

...effective instructors!

Quality control

Some licensed professional pilots should never have been given their jobs, and no amount of training will rectify this situation! So say the US General Accounting Office (GAO), in a report published in June 1996. After investigating several recent fatal accidents, it has reported its serious concern about the pilots' poor standards of knowledge of the aircraft systems. It concludes that internationally accepted minimum standards for pilot training had fallen behind developments in the air transport industry. Pilot proficiency failure was the cause 58% of flightcrew human factor related accidents and this upward spiral continues.

Under the heading 'Selected aviation incidents illustrating human factor issues', the GAO lists brief examples of real events – 'the pilot did not achieve a satisfactory level of performance, despite remedial training; the aeroplane crashed because the pilot did not take corrective action, even though the first officer advised him appropriately; the captain's previous employment had been terminated because of inadequate performance, but he had been re-hired by another airline; etc.'

The task of assessing potential pilot employees for suitability, and of carrying out their recurrent training and competency checks, is largely devolved to the airlines by national aviation authorities everywhere. Unfortunately, flightcrew licensing systems, airline selection procedures and assessment during initial type rating/recurrent training have in the past failed to filter out these poor quality pilots and not only let them gain access to flying jobs in commercial aviation but also to stay there.

Selection

Despite the failures mentioned above, most pilots who reach line flying are good enough and effective re-current training sets out to keep them that way. The key to weeding out these undesirable pilots is in good selection during the initial assessment phase. Experience has indicated that if someone shows marginal performance, even during ab-initio training, then this will be reflected in line operations. If selection has failed to ensure that pilots have aptitude then it is a pointless exercise continuing.

Standards and safety, theoretically, govern the decision about whether to hire a pilot, or keep a line pilot flying when, for example his performance remains marginal after remedial training. The decisions *should* be independent of other considerations, such as aircrew shortages (or surpluses). However, the recent record growth of the airline industry world-wide has led to a significant decrease of qualified pilot candidates able to fill the vacancies within the ranks of airlines, especially the regionals. To keep up with demand for pilots, regional airlines have been seen to lower their minimum hiring requirements. The fear is that a continued hiring boom will inevitably lead to an overall lowering of standards.

Lack of experience, inherent lack of aptitude, lack of knowledge or poor training are all problems that can be addressed. It is the job not only of the licensing regulators, the ab-initio training schools, the airline selection boards, but also it is primarily in the airline training department where these issues can be addressed.

Management styles

There is a perceived difference between airline management styles. Most pilots when choosing an employer are aware of the differences offered in salary and conditions, lifestyle, operation, training standards and status. The levels of skill and training offered by an individual mean the best pilots go for the best jobs. At a time of industry expansion, a highly skilled and experienced pilot is well placed to choose a company that offers the best package. However, with the fluctuating laws of supply and demand it is not always possible to select the company of your choice.

Flight crew training represents a large slice of a company's budget and pressures are to cut costs. Airlines are being forced to analyse ways to examine all the issues associated with employing pilots. Right at this moment, certainly within Europe and the USA, there is an increasing demand, with a dwindling supply, largely due to the drop in numbers of ex-military trained pilots and the phasing out of the 'self-improver' route. Apart from ab-initio company sponsored training schemes, a pilot wishing to set out on a career will have to fund his own training to basic licence standard and then go shopping in the uncertain market place for a job.

The past decade has seen the emergence of many so called 'new style' airlines whose management style differs from that of the 'traditional' airline. The 'new style' airline economises on salaries and benefits, does not invest in a training infrastructure, demands high productivity, accepts low crew experience and does not use a selection procedure beyond licence qualifications and interview and has a high attrition rate. The 'traditional airline' by comparison offers better salary and benefits, has its own training infrastructure and a thorough selection procedure but understandably has a higher cost base. The 'traditional' airline tends to offer a better lifestyle, with a perceived higher quality of company culture, having long serving and experienced crews whose skills are generally above the pilot average.

Predictably, the most able pilots will, given the opportunity, gravitate towards the 'traditional' airline for a career. Statistics have shown that for each vacancy a 'new style' airline will receive 3 applicants where the 'traditional' airline will receive 35. The implications here for long term overall quality of flight safety for these 'new style' airlines are an industry problem and one that will need to be addressed by the training pilot.

Guide to training pilot assessment

Positive	*Negative*

Personality

Positive	Negative
Imaginative	Stumbling
Confident	Defeatist
Commanding	Hesitant
Patient	Resentful
Knowledgeable	Bluffing
Flexible	Abrupt

Enthusiastic	Dull
Encouraging	Intense
Humorous	Dry
Lively	Distracting mannerisms

Voice

Fluent	Monotonous
Audible	Verbose
Concise	Poor articulation
Precision	Incorrect phrasing
Variation	Disjointed
Inflexion	Obscure

Direction

Emphatic	Irrelevant
Relevant	Rambling
Concise	Too long
Stimulating	Routine
Aims clearly stated	Poor linking

Production

Logical	Inaccurate
Interest maintained	Protracted
Variation	Poor timing
Emphasis	Digressions
Recap/recall	Hasty
Well prepared	Protracted

Participation

Controlled	Irrelevant
Questioning	Dictating
Timing	Poor logical control
Body language	Bad timing
Optimistic	Complicated
Well planned	Obscure

It should be remembered that elements of instructional technique are very difficult to isolate and describe. All humans are accustomed to giving explanation, guiding, demonstrating and showing others how to do things. When an attempt is made to turn these natural capabilities into conscious, professional skills, their abstract qualities can make the task difficult.

…instructional technique

To communicate about instruction, some type of behaviour category description is useful. This provides a means of analysing and discussing instructional communication, which can be used in the description of a trainee's performance, to control and enhance instructional skills.

- Giving information
- Demonstrating
- Giving direction
- Giving instruction
- Cueing
- Testing
- Positive evaluation
- Seeking information
- Negative evaluation
- Defend/attack
- Illustrating
- Integrating
- Structuring

- Emphasizing
- Summarizing
- Encouraging
- Facilitating
- Consolidating
- Questioning
- Reacting
- Repeating
- Enthusing
- Drive/draw style
- Giving asides
- Creating a climate
- Body language
- Personality.

A fundamental problem facing all airline training pilots is the human tendency to form a *mental set*. A presumption so strong that it affects the whole thinking process. It has a quality that enables each person to recognize pre-conceived conditions, when they arise, and respond quickly. However, it can lead to ineffective handling of events, which are not quite as anticipated. This *mental set* is caused by:

- Implacable opinion
- Value judgements
- Egotism
- Intentional orientation
- Personality.

It can lead directly to an imbalanced amount of perceptive perception, which can disrupt instruction, and is a frequent cause of *personality clash* between trainer and trainee.

...implacable opinion!

Leadership

The term *leadership* is capable of so many widely differing interpretations that it tends to be a stumbling block to the understanding of many people. Leadership is fundamental to the effectiveness of many jobs, but especially so in the dynamic airline environment. Some leadership styles are easily identifiable:

- Leadership by example. A force pulling from the front. If this type of leader is himself highly skilled and competent he may be so far removed from the level of skill of his trainee that he his unaware of, or unsympathetic to, the difficulties his trainee may be experiencing. Being at the front, he will tend to take all the decisions all the time. The trainee therefore will not be involved in the experience as much as they might be, tagging along without any real idea of what is going on.

- Leadership from the rear. A force pushing from the back. The danger here is that the trainee may suffer from loss of direction, and become unsure of what he is trying to achieve. If pushed too hard from the rear then anger and frustration can set in. Careful discretion needs to be applied if using this type of leadership style.

- Leadership from the centre. This is symbolic of the modern sociological concept of leadership. Such a leader is a combination of both styles mentioned above. He will only be authoritarian and act in a military style when the situation demands. During a training session, he will be at the front, the back or the centre when occasion or situation demands. He will work democratically as far as possible, helping not dictating, to his trainee to resolve their difficulties. He will be watchful to give support, encouragement, sympathy, guidance or advice. He will be tuned in to signs of discomfort, stress or anxiety. He will be alive to possibilities that will arouse interest and enthusiasm. He will act as consultant, counsellor, guide, and mentor and be a source of information, knowledge, skill and experience.

Assessing leadership skills during pilot selection may be difficult, but should form an important part of the selection process.

Leadership skills

- As a trainer, one of your most valuable aids is the ability to recall past experience, particularly how you felt as a trainee. It will enable you to develop insight, a valuable tool.

- Planning, forethought and anticipation are skills, which will enable you to meet difficulties in a state of greater preparedness and provide you with a ready course of action should emergencies arise.

- Habit of checking. What have I done? What haven't I done? What else can I do? With practice, this sixth sense is always switched on.

- Habit of observation. Supplementary to the habit of checking. The ability to perceive things varies greatly in people, it is mainly by observation that you will get to know the characteristics, strengths and weaknesses of your trainees. The non-verbal signals of body language will tell you much of what you might like to know in terms of tiredness, boredom, low morale, tension, anxiety, stress, personality conflicts, feelings of insecurity etc.

- Decision making. A wise leader will contrive to get his group participating in as many decisions as he can safely allow.

- Learn about your own limitations, get to know yourself. You need to have examined within yourself your reasons and motives, for taking on this task as a trainer. You need to have clarified for yourself why you want to do it at all, and to what ends. There needs to be a clear realisation that the responsibilities of leadership impose a discipline that allows no room for your own wishes, ambitions or aspirations.

Motivation

Nearly all airline pilots are strongly motivated in their chosen career. This makes the task of pilot selection a good deal easier than in some other professions. However, practical experience has shown that the trainer needs to provide recognition, responsibility and work that is challenging. As a training captain, it is all too easy to become a big de-motivator to the trainee, if he perceives he is being unfairly assessed.

Motivation is vital to any job if people are to give their best. People are undoubtedly a most critical resource, and no matter what the degree of sophistication that is attained with advanced technology, so much still depends on the *human factor*. It is the task of the training captain not only to teach and assess the specialized task of flying and operating a particular aircraft, but also to further motivate his trainee, by creating the best environment for the trainee to learn and give the best of his work. Motivation is what makes people do things. It is what makes them put that extra effort and energy into what they do. It does vary both in nature and intensity, from individual to individual, and from day to day, depending on the particular chemistry at any given moment.

- Positive motivation occurs when people give in to a request

- Motivation ceases when people are compelled to surrender to a demand.

...getting to do it willingly!

Signs of motivation

- High performance and results being consistently achieved
- The energy, enthusiasm and determination to succeed
- Unstinting co-operation in overcoming problems
- Willingness to accept responsibility
- Willingness to accept change.

Lack of motivation

- Apathy and indifference to the job
- Poor timekeeping and high absenteeism
- Exaggeration of problems, disputes and grievances
- Lack of co-operation in dealing with problems and difficulties
- Resistance to change.

Trainees need to know their tasks, what is expected of them and to what standard, if they are to apply their efforts usefully. They need to know how well they are doing, in order to create a feeling of achievement. They need to know what policies or changes have been decided that will affect them, and they need to understand the reasons for these decisions. They also need the opportunity to contribute their own ideas to instil a feeling of job satisfaction, making the overall organization more effective. The base training captain is an important link in achieving these objectives. Failure to communicate in any management situation is costly. Time, co-operation, commitment, morale and overall effectiveness are all at risk.

It has been suggested that there are four kinds of people in the world:

- People who watch things happen
- People to whom things happen
- People who do not know what is happening
- People who make things happen.

...do not know what is happening!

Effectiveness

Research has shown that the most effective training pilots are the ones that use the following behaviours:

- Integrating
- Structuring
- Emphazising
- Illustrating
- Cueing
- Seeking information
- Testing learning
- Giving instruction
- Giving explanation
- Positive evaluation.

Ineffectiveness

- Absence of testing learning
- Defend/attack
- Negative evaluation
- Creating harsh, aggressive, instructional climate.

In a training environment, when all reaction is withdrawn, there is usually a feeling of a hostile situation. This creates a harsh, unpleasant climate. Rapport is lost and a fresh start becomes difficult. A training pilot *has* to react, and both positive and negative evaluations have their appropriate moments. Instructional climate will be judged on the balance of these behaviours. Low reaction is symptomatic of partially broken down communication and is particularly ineffective and unpleasant.

Style

A flexibility of instructional style should be highlighted as a crucial factor in the effectiveness and economy of instruction:

- Drive - directive, authoritative, one way
- Draw - inductive, encouraging, essentially two way.

Good instructors will use both of these styles to suit appropriate moments. Airline training pilots should be first class aviators, yet it must be remembered that the best pilots are not necessarily the best instructors. The key to success in airline flight training, is the adoption of a high degree of professionalism, in the special forms of communication, which help people learn.

The training pilot - selection and suitability

Some pilots were asked to comment on their trainers from the point of view of being:

- Adequately prepared
- Properly motivated
- Suitably selected
- In need of further guidance.

Most pilots questioned were well satisfied with their trainers from the point of view of their motivation, preparation and effort exhibited by individuals, in carrying out the task. However a significant number felt that selection methods could be improved, and that there were a few people in the training role who should not be there. Some pilots were critical of the attitude displayed by their trainer, in that he played a *safety pilot* role rather than an active training role. One first officer claimed that he learned more about landing the aircraft from the cabin staff during a *plateau* problem of bad landings. The trainer, although a pleasant and amicable personality could only *flinch* at the repeated heavy landings and say: *Don't worry, you'll soon get the hang of it.* Similar comments about training pilot suitability were made in varying degrees by other pilots. Some trainers were cited as good pilots but poor teachers. Others felt that the role of the training pilot was not sufficiently defined. Asked to comment on improvements to the system the following suggestions were made:

- Careful selection of pilots for the training role, using grounds of suitability as well as seniority

- Improved guidance to newly appointed training pilots, with particular emphasis on the theory of learning and practical instrumentation considerations. Most felt that a flight training manual was a positive step in this direction

- An improved training pilot *school*, with more emphasis on a practical approach. Activities should include discussion, practical briefing, and mutual exchange of case histories. Experienced trainers from various aircraft types should be available to speak on particular subjects and answer questions

- More frequent type refresher courses should be run

- Selected trainers should be given clearance to operate and fly the simulator for practice sessions, as opposed to training sessions, during the currency of line training.

Training programmes

- Trainers were asked whether they had devised a programme of training:

- Other than the training form itself, or

- Should the training form be arranged as a sequential *lesson* programme, or

- Is it better to operate on a day to day basis, according to the needs of the trainee, with very little formal input?

In answer to these questions, most trainers indicated that they thought:

- The present training form was generally adequate to indicate coverage of material but should not be used as a programme.

- Flexibility, was the chief requirement, largely due to roster changes. The training pilot should know the training potential of each trip or flight cycle for his type of aircraft, so that training exercises could be planned with maximum benefits to the trainee and minimum delay to the schedule.

- It was generally desirable to devise a training plan in order to best integrate theoretical learning and operating manual study with practical flying. Properly thought out, this programme would provide repetition and practice and avoid a last minute rush to cover all the required subject matter.

- The programme depended on time being set aside for pre-flight briefing and discussion.

Advanced cockpit technology

Trainers were asked their thoughts on training when using advanced cockpit technology. Is it better to use a method of:

- Basic aircraft first, then gradual introduction of the automation? or
- A fully integrated approach to give maximum exposure to automatics?

This question led to a great deal of discussion and hotly debated ideas. The majority favoured the first case, but both groups felt that it was most important not to over concentrate on the use of automation as it would inevitably result in a degradation of both manipulative skills, and support and monitoring duties. It was also felt that trainers should place particular emphasis on showing how the automation helps the flight and reduces workload, rather than presenting a *gimmick* approach. A balance must be maintained during the training period so that instrument flying skills are maintained. Trainers further indicated that intelligent use of the autothrottle was particularly important. A dogmatic approach should be avoided. The trainee should be encouraged to experiment with different auto-modes but should never lose the feeling of being in complete control of the aircraft.

Using the automatics in coupled approach situations should receive more emphasis than endeavouring to leave the autopilot engaged throughout a complete visual approach. The visual approach should be used to fly manual exercises wherever possible, particularly early in the training, to get a basic *feel* for the aircraft and to improve judgement. Care must be taken to avoid unwanted power changes and spin up by autothrottle during the descent. This practice is costly, time consuming and detrimental to passenger comfort.

A study by the US FAA and National Aeronautics and Space Administration (NASA), found that pilots using modern automated cockpits have to cope with an excessive workload, and perform worse than counterparts flying in aircraft fitted with old technology. Although this research now is somewhat dated, the design and training for glass cockpits *has* improved, many of the points it raises are still valid.[1]

During this study, airline crews took part in a specially designed simulated flight in which 12 two man crews of a traditional DC9-32 were compared to 12 two man crews of a MD80 *glass cockpit*. Information also came from 84 pilots who gave details of flying experience. A further 73 filled in questionnaires. The performance differences were generally small, but always favoured the DC-9. Overall, performance of the DC-9 crews were superior to the MD-80. It has long been the dream of traditional engineers to automate human error out of the system - *we found little to comfort those who see automation as a cure for human error.*

Research found that when the workload was high, the automation increased it. A growing number of accidents have been linked to pilots failing to interact with the automated equipment. Problems occur by the pilots having too much choice. The report calls for better training and for glass cockpits to be more *user friendly*.

When training in the use of advanced cockpit technology, it is better to use a method of basic aircraft first, then gradual introduction of automation. It is important not to over concentrate on the use of automation as it not only results in a degradation of manipulative skills, support and monitoring duties but can increase the pilot's workload. A balance must be maintained so that flying skills are not impaired. Emphasis should be placed on showing how the automation helps the flight and reduces the workload. That *glass cockpits* are here to stay is undeniable. What we as pilots must do is understand the limitations they impose upon us if they are used indiscriminately.

Overcoming the training plateaux

Trainers were asked what techniques they had found to be successful in helping a trainee at times when he appeared to be not making progress, or even regressing.

---

[1]The Impact of Cockpit Automation on Crew Co-ordination and Communication, US FAA and National Aeronautics and Space Administration 1989

...relinquish control to computers!

The following suggestions were made:

- Abandon the new learning task and let him go back to something which he knows he can do well.
- Ease up the training pressure. Fly the aircraft yourself and let your trainee concentrate on support duties.
- Change the subject completely, or approach the same subject from a different angle. Do not let him persevere with repeated unsuccessful attempts.
- Introduce a third or fourth person to have a discussion on the subject. This is particularly successful if done away from the aircraft on a social basis. Let him see that he is not the only one to have experienced the problem. Use humour to help regain perspective.

Psychometric assessment

- Psychologically based training techniques, like crew resource management (CRM) and total qualification programmes (TQP), *(see Chapter 14 TQP)*, are becoming more widely used, and basic human factors knowledge is now part of the licence examination requirement for pilots

- The existing system of regular pilot assessment by training pilots, plus the fact that cockpit voice recorders (CVRs) and flight data recorders (FDRs) are installed throughout the airline fleet, make quick access of performance analysis

- An additional role for psychometric analysis for line pilots has yet to be defined.

Psychometric testing is more appropriate to screening ab-initio trainees for airline sponsored training. In any event, many of the so-called objective testing methods, are themselves subject to dubious qualities of biased evaluation and poor validation when directed to individuals.

Personality as a predictor of attitude and performance

Crew effectiveness in multi-crew aircraft is determined by technical skills, attitudes, behaviour and personality characteristics. Traditional training has largely centred on technical proficiency. Unlike technical skills and, to some extent, attitudes, personality is not likely to be changed by specialized training. While personality probably cannot, and certainly should not, attempted to be changed, a person's behaviour, while undertaking professional duties, is of legitimate concern.

...psychometric testing!

Personality traits are resistant to change and there is evidence that *some* personality characteristics are particularly undesirable in a multi crew, total aviation resource management (ARM) environment. The importance of *selecting in* the right personnel, both for the airline training department and for line pilots, is an area that perhaps in the past has not received the attention it deserves. Relevant personality traits *should* be a focus of selection.

For a number of years, a team,[1] sponsored by the NASA-Ames Research Center, at the Department of Psychology, University of Texas at Austin, have been working on the isolation of personality factors, related to performance and adjustment in demanding environments, including aviation and space operations. Their findings centred around a personal characteristics inventory (PCI), and cockpit management attitudes questionnaire (CMAQ), which was designed to identify underlying trait dimensions. These broadly consisted of goal orientation or instrumentality, and interpersonal orientation or expressivity. These categories were represented by both positive and negative components.

Goal orientation - instrumentality

- Achievement motivation
- Mastery - an interest in undertaking challenging tasks
- Work - desire and satisfaction gained from working hard
- Competitiveness - desire to do well and beat others in activities.

Interpersonal orientation - expressivity

- Warmth
- Sensitivity
- Awareness
- Willingness.

Negative instrumentality was measured as a subscale, which reflected autocratic, dictatorial orientation. Negative expressivity was measured, reflecting verbal aggression and a nagging hostility directed towards others.

- *'The right stuff'*

Above average levels of instrumentality, expressivity, mastery and work and

---

[1]Validating Personality Constructs for Pilot Selection - Robert L. Helmreich and John A. Wilhelm - Dept. of Psychology, University of Texas at Austin. 1989

below average levels on negative instrumentality and verbal aggression. This combination of characteristics is considered optimal for the close co-ordination required by crewmembers in multi-crew aircraft.

- *'The wrong stuff'*

Above average levels of instrumentality, negative instrumentality and verbal aggression, as well as work, mastery and competitiveness.

- *'The no stuff'*

Below average on instrumentality, expressiveness, mastery, work and competitiveness.

- *'The arrogant and unmotivated'*

High in negative instrumentality and verbal aggression, low in both instrumentality and expressivity and very low on achievement scales.

Sub-populations

The results suggest that meaningful sub-populations exist amongst pilots, and that these groupings are useful in understanding how personality factors act as variables in the determination of attitudes and of performance. The *right stuff* are obviously going to show the more desirable patterns of responses, and should be *selected in*, whereas the other groups are less desirable and should be *selected out*.

A study conducted in a B737 simulator at the NASA-Ames Research Center, provided important findings, regarding personality and performance and the limitations of training.

Volunteer crewmembers were tested on the PCI and crews were formed on the basis of personality. The crews then flew a two-day, five-leg simulation that involved significant mechanical abnormalities and marginal weather.

Extensive data collected showed that the most effective crews were led by captains with personalities from the *right stuff* cluster, whilst the least effective crews had *no stuff* leaders. An interesting pattern emerged in crews led by captains with the autocratic, domineering, *wrong stuff* cluster. These crews performed worst during the abnormal situation on the first day, but their performance was much improved on the second day, suggesting that fellow crewmembers may have learned to adapt to this difficult type of leader over a period.

The cultural effect

Aviation by its nature is international and based on common precepts that hitherto have been predominately western and individualistic. Yet an increasing proportion of the world's airline pilots are not from western cultures but from cultures more passive and respectful in nature. *(see Chapter 13 Culture & CRM)*

When screening applicants, cultural differences should be acknowledged and not be taken as the 'wrong stuff'. Selectors should be wary of their own bias and prejudice and be objective throughout any screening procedure of pilot applicants.

A strategy for screening applicants

There is evidence that personality screening of pilot applicants can make a valuable contribution to the selection process, providing the criteria is long term line performance rather than performance in initial training. The PCI measures attributes that have been validated as determining crew performance. Personality measurements, such as the PCI, are however unrelated to measurements of intelligence, aptitude, and psychomotor skills. Screening in this area is obviously also important. There are a number of effective IQ and multidimensional aptitude tests available in this area. Psychomotor skills can be tested by a short session in a flight simulator, giving useful data as to how an individual performs under pressure.

Applicants who receive good ratings in interviews, by being articulate, expressing enthusiasm and motivation for a position, fit the classic profile, but are not necessarily the *right stuff.* Objective testing is a necessary companion.

Under the recently adopted European Joint Airworthiness Requirements (JAR), flight crew licensing have now implemented a formalised structure for the crucial parts of ab-initio training for self sponsored pilots undertaking commercial pilot training. This has effectively killed off the 'self-improver' route, which largely relied on accumulating hours by whatever means possible, then cramming for examinations. This has been designed to improve overall standards of commercial pilots. The effect however, could be the cause of a pilot shortage, as there will be fewer pilots in the system than before due to the cost of a formalised course. The supply of military pilots, a traditional, formally trained source for the airlines, is drying up and many, even well established airlines do not have a philosophy of training their own from ab-initio stage.

Pilot career appraisal and development

A number of airlines have adopted a career appraisal development scheme (CAD) designed to review the pilot's progress within the company. Ideally, any such scheme will give individuals the opportunity to:

- Consider their own training and development needs
- Assess their own performance
- Discuss how they can maintain and improve their performance
- Clarify and agree their key objectives
- Agree their training and development needs.

The company benefits from such a scheme by:

- Creating a more motivated and directed workforce
- Identifying people with potential
- Identifying the training and development needs of each pilot
- Acquiring a method to develop an *in-house* total qualification programme.

With more emphasis on career structure and standards, more information is required on pilot performance, including flying ability, initiative and attitude, airmanship, crew co-operation, aircraft management, command ability or potential, technical knowledge and general aviation resource management. The CAD scheme introduces a systematic and quantitative analysis of individual pilot progress and development, and forms the starting point for any company wishing to develop its own total qualification programme (TQP).

The next chapter investigates training trainers. This is an area that has, in the past, had very little attention.

Experience has indicated that if someone shows marginal performance, even during ab-initio training then this will be reflected in future line operations.

12 Training trainers

Training trainers is a subject on which many people have numerous theories. In recent times, much literature has been produced in the field of business training but very little, if any, for the airline trainer. Budding airline trainers are not usually able to find much directed at their particular subject. However, by searching various other fields such as teacher training, military training, and business training, one can find that many books contain the same subject matter and very similar theories. Does this surprise you? It should not, because training is just dealing with people. You are a person, so you should know what you like, and what you do not.

Consider a statement from Sir Winston Churchill, the British second world war leader - *personally, I am always ready to learn, although I am not always ready to be taught*. This statement encapsulates the basic task of teaching, in that, as a definition we could say that teaching is *creating an environment in which someone is able to learn*.

Churchill was stating the obvious. If one came to him and produced a theoretical lesson, presented with no personality or consideration for the trainee's thoughts and feelings and existing level of knowledge, one would soon have a very bored or even irate trainee. People usually come to the airline training environment in adult life and have a positive attitude towards the tasks at hand. If you as a trainer are sensitive to this and want to make your instruction interesting then you will succeed.

Background

Before going any further, let us consider where our potential trainers are coming from. Well usually, from the line pilots. But where have they come from? What do they do or have they done in the past? Some may have been officers in the military *(trainers?)*, some may have been in a sports club *(trainers?)*, and some may have been teachers before starting flying. Most probably, some have children *(trainers?)*. To get you started here you can see that most people will have had some sort of experience, no matter how seemingly irrelevant, in training people. The principles are the same. Now try a simple experiment. Take two subjects, one, which you know intimately, get someone to allocate the other. Take fifteen minutes to prepare each one for a three-minute presentation. Ask your long-suffering partner or friend which was the better presentation. It is more likely that the known subject will have been better. The moral of the story is, if you know what you are talking about, then the message will come across easily. This is the message you have to impart to your trainee trainers. Now how do we train trainers?

...know what you are talking about!

Why

We realize that many of our pilots have had some sort of training background so why do we need a training course? The answer is on the one hand to standardize the method of instruction in the company, and the other is to get trainers used to discussing methods and to learn from each other so that the basic standard of instruction is kept at a high level.

How

A company or aviation authority should consider having three levels of course, the basic course, the advanced course and the refresher. After a thorough selection process to choose the most suitable candidates for training, everyone should attend a basic course unless they have been training for years and this may be unproductive. The experienced trainer should attend the advanced course before starting training, to learn the company philosophy.

The selection process is so very important, not only to get the right candidate but because it can be damaging to the training departments' organizational culture to have someone who is unsuitable. One usually has hearsay and sometimes trainee reports. Usually the trainees are reluctant to complain about a trainer for fear that it will affect their own careers. The training courses should be run in a logical sequence and within a reasonable time frame. A course lasting five days would be considered normal.

...is this going to affect my career!

Basic course

The basic course should contain as much practical work as possible otherwise your trainee trainers will not be content. Ask yourself what they expect from such a course and you will realise that they actually want to know how to walk into the room and the simulator and start instructing. They also want to know how to deal with all the different types of trainees. They also want to know how to recognize all the different types of trainee. In short, your trainee trainers want to be injected with thirty years of instructional experience in five days. They do not want to make fools of themselves in front of their trainees!

To achieve this passage of experience is simply impossible. What we can achieve is to open our potential trainer's eyes. Show them what works and what does not, and what theories are currently in trend. Get them to discuss each subject and aspect. Once they have started to discuss the theories, they will begin to build up some confidence leading them to relax. They will then realize that training is not an obscure science, but something that everyone can do if they are willing to put some effort into it.

What is a very important part of the basic course, is to have your trainee trainers prepare short presentations both individual and group work. These should be videoed, played back to the class and discussed. Everyone can then have the joy of see themselves as others see them. Perhaps they will want to give up at this stage! The group work presentations are good because the trainee trainers will gain insight into how others prepare themselves. As a suggestion for general content of the course, the following subjects should be considered:

- Teaching theory
- Learning theory
- Presentation skill - theory and practical
- Briefing skills - theory and practical
- Debriefing skills - theory and practical
- Discussion on assessment and qualification
- Report writing
- Real case studies - discussion
- Presentation from existing trainers as examples
- Discussion with company managers / directors
- Social aspects - end of course get together!

The course should be intensive, so that the trainees have to work under *some* pressure. Instructing has to be done under stressful conditions in the airline environment and there is little room for people who can not maintain a steady humour under stress. However, the aim of any training must be to offer optimum conditions in which people can learn. The job of the trainer should be to create a suitable learning environment by being aware of the psychological state of their trainee. The principle object is to produce an effective trainer by allowing him to first learn the requisite knowledge and skills. There should be ample opportunity for social interaction, having a drink together or a meal, so that the discussions can develop out of the course room. After such a comprehensive course, the trainee trainer will find that although he has little more instructional experience than before, he will feel more confident.

Trainer selection

Selecting trainers that have already demonstrated their knowledge, skills and sound attitudes will minimise the use of training resources, reduce the possibility of failures and will prevent the wrong message being passed down to crews, affecting the whole training department's credibility. In practice, competence in behavioural skills training requires considerable experience. Demonstrated basic skills to a high standard would be a minimum requirement for trainer selection. As a starting point, applying the following criteria could assess basic skill knowledge:

- Personality types
- Perceptions
- Self awareness – strengths and weaknesses – motivation
- Thought processes
- Behaviours – their differences and effects
- Team building
- Problem solving and decision making
- Human factors and limitations
- Understanding roles – leadership etc.
- Error management
- Ergonomics and effects of automation
- Understanding of individual and cultural differences
- Conflict management styles
- Physiological conditions – stress, fatigue etc
- Relevance of crew resource management (CRM)
- Communication
- Active and passive listening
- Questioning – checking – understanding
- Reading body language
- Demonstrating empathy
- Giving/receiving criticism and praise constructively
- Information processing
- Coaching
- Patience
- Situation awareness
- Delegating
- Prioritizing
- Etc.

The list is endless and needs in addition to address the values and beliefs that influence people to select a set behaviour. Commitment, professionalism, motivation, attention to detail, respecting their own rights and the rights of others, being willing to take responsibility for decisions are all deciding factors in the selection process.

Trainer competency standards

As a guideline to achieving the training objective, the trainer should attempt to match the competency standards currently been set by some of the major operators:

- Trainers should deliver training in a manner that is respectful, participative and open
- Demonstrate approachability, calmness and self control
- Develop new approaches and improved product services without being constrained by past experience or current practice
- Demonstrate and role model an assertive communication style
- Challenge and draw out individual feelings and opinions
- Be able to give and receive constructive criticism
- Be able to distinguish between process and content
- Demonstrate effective communication and interpersonal skills
- Encourage open and interactive discussion
- Be patient and constructive in probing into areas where improvement is needed
- Ensure that all crew members participate in the discussion, effectively drawing out quiet or hostile crew members
- Provide a clear summary of key learning points.

Becoming a competent trainer

The evaluation of training quality is a complex issue, as competence in behavioural skills training requires considerable experience and dedication. Positive flight crew reactions to training are essential for training effectiveness. If crews' perceptions of quality are low, or a threat, or feel that it is not relevant then they are unlikely to benefit from the experience. How many pilots *still* go for their recurrent training (having found out from rostering who their trainer is) with fear and dread. Judgement of the quality of training can be ascribed as follows:

- Professional credibility. Flight crews, perhaps more than most professions are continuously scrutinised for professional competence and they demand high standards from those who would presume to teach them.
- Practical relevance. Training input that is not perceived to be directly linked to the job is distracting, time consuming and a big turn off.
- Amusement value. A memorable lesson in a busy, stimulating and exciting environment is an essential element to counteract highly factual training.
- Self-development. All pilots are keen to learn, improve their job skills and most are highly motivated. You as a trainer might be judging them but be assured that they too are judging you.

Advanced training

A recommended programme to address CRM/LOFT facilitation skills would typically involve the participant attending a CRM skills course (3 days) followed by facilitator course to develop facilitation skills and the ability to manage difficult situations (4/5 days). The trainer then would shadow an experienced trainer on a skills course who then facilitates the course with support from experienced trainer. The trainer is then finally assessed as competent (or otherwise) to facilitate either the whole skills course or elements of it. Activities on a facilitator course would emphasise the following:

- Understanding learning styles and group dynamics
- Developing skills to a high standard through case study and role play
- Understanding conflict and developing skills to manage difficult situations
- Developing counselling skills
- Reviewing competency standards and regulations
- Developing personal confidence and awareness
- Evaluation techniques at all levels
- Understanding the learning process and environment.

An essential exercise is for trainers to regularly get together to discuss relevant topics.   It is a forum where everyone can push his or her suggestions and ideas. Individual training problems can be aired. Everyone will have differing experiences and the exchange will then fall on experienced ears. It may be that a similar thing happened to another trainer but he was able to deal with it, so the outcome was positive, whereas the other may not have been so successful. The exchange of methods will then cut corners for the future.

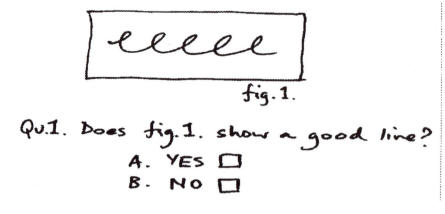

fig. 1.

Qu.1. Does fig.1. show a good line?
A. YES □
B. NO □

...the line check!

Flight deck management training (FDM)

Every flight is characterized by constant change. Although the laws of physics precisely define the flight envelope, the 'pilot management' envelope is subject to many variables. Flight deck management training should aim to expand the pilot's skill, not only in developing and maintaining his individual proficiency in flying skills but enable him to operate effectively as part of a team.

The crew's use of its flight deck management skills starts with the development of a plan for conduct of the flight, requiring sound knowledge and application of standard operating procedures, as well as company policies and regulations. The aim is to ensure that the whole crew share a common image of how the flight will be conducted. Challenges, as the flight progresses requires various skills as the crew monitors events and communicates with the others the desired response. With time-critical challenges the optimum response is the conservative one, that best moves the aircraft or crew away from the edge of the envelope.

As we have discussed elsewhere in this book, the majority of accidents/incidents are the result of poor performance by the crew in several subsequent series of cycles and actions. Application of flight deck management training, during train the trainer courses, within the organizational company culture, is a desirable means of teaching 'error management' and the philosophy of pilot-preventable accidents.

Training trainers in FDM can be achieved in the classroom by having the 'crew' participate in an interactive discussion of a representative mission scenario. The simulator will provide scenarios that are more realistic provided Line Orientated Simulation (LOS) methods are applied. Here the instructor acts as facilitator to maintain the realism of the flight. By acting as facilitator, not interrupting during the session, using video replay and angling the de-briefing so that the candidates de-brief themselves, plays a key role in effective training. *(see Chapter 14, Total qualification programme)*

- Tell me and I'll forget
- Show me and I'll remember, but
- Involve me and I'll understand.

Refresher

Refresher training for all trainers is an essential step in maintaining healthy, updated and motivated training staff. Recurrent skills training should be delivered in a modular form, on a regular basis and over a period. The skills base is thus broadened and gradually improved. The course should take trainers with differing levels of experience, in order that the most experienced may transfer to the less experienced and vice-versa. Vice-versa, you may ask why? Well, cases have been recorded where an experienced instructor was losing motivation having taught for so long. A younger colleague gave a presentation on motivation and how he addressed the problem, this pumped some new life into the older trainer which both very useful and satisfying. There have been occasions, where trainers of many years' experience have made presentations to colleagues and it was a real disaster. It usually turned out that they had never been taught to train, and had thought that what they were doing was all right because they had had no benchmark for comparison.

The refresher course should last for two days and should follow normal working hours. You will find that two days is necessary as people do rethink things overnight. The second day then gives an opportunity for more discussion. Short presentations should also form part of the course. People always learn from each other.

...a disastrous presentation!

Commonality

Obviously, airlines are in business to offer a service and make money. Training within airlines is also a profit centre for those companies who have the spare capacity and facilities to offer to others. One has to remember that all aircraft use the same airspace, therefore it is desirable for the companies who have facilities to ensure that those who have not maintain a reasonable standard. In the area of training trainers and human factors there now seems to be a closer liaison and exchange of views. This can only be for the advantage of the industry in general, and is a very real boost to improving overall flight safety.

Sharing experience in training should include techniques, knowledge gained from line experience, and even instructors. There should be no problem with exchange of trainers to observe others at work. This can only bring new ideas and improvement all round. What is definitely wrong is to mix training. A trainee should attend a full training programme within one system before moving about. The reason for this is to try to obtain a standardized product and not confuse the trainee or exclude certain areas unintentionally.

When exchanging trainers one must make sure that the company training plan is not interrupted and that the company trainees do not suffer in any way by, for example, being forced to have observers in the simulators during check rides.

...sharing ideas!

Standardization

We have talked about standardization within the airline itself, but standardization within the industry is also a concept as mentioned in the previous subsection. In smaller companies, or countries with fewer resources, standardized training of trainers should be considered for all aviation trainers. By this, one can ensure that a potential airline or military trainer has done the same basic and advanced courses. There will of course be a need for specialized training in the field in which the trainers will work, ab-initio, airline, advanced military and so forth. By constructing a common training programme, those who have to foot the bill together with national standards and flight safety will see a lasting long term benefit.

Summary

Training trainers is done in all branches in industry and government. Most of the basic aspects are applicable to all areas whether flying, military, young children or business. Courses can easily be tailored to suit specific requirements and can be given by newly trained trainers. Some hard work and imagination are required and a good course can be the result. It is also imaginable that the basic elements of training can be common to all branches within the company.

The next chapter continues the investigation into culture and crew resource management.

Operational training briefing.

13 Culture and CRM

In recent years the pressures and demands on pilots have changed as aircraft have become more complex and airspace more crowded. Although advances in aeronautical engineering and avionics have made a considerable contribution to flight safety, they have also raised new questions about how people interact both with technology and each other. Increasing our knowledge and understanding of the operational and human factor elements involved are seen as the key to further improvements.

The need for Crew Resource Management (CRM) training is now well established as an integral part of crew training and aircraft operation. There is general support for CRM training but concern that there is at present a wide variation in standard and consistency, perhaps allowing undesirable practices to perpetuate. It is an undeniable fact that CRM standards of performance have a bearing on flight safety and the efficiency of aircraft operations. Being essentially more explicit and refined than the professional standards implicit in the term 'airmanship', CRM is a long-term development process that encompasses a variety of resources, running from the traditional to the highly interactive and experiential learning. As in any other field of training knowledge, ability, and motivation are all necessary to effect enduring changes in behaviour.

CRM! What is it?

Most CRM training in the past has focussed on the avoidance of error. A shift in training philosophy now leans CRM training to be more accented towards error countermeasures, focussing on achieving knowledge, skill and attitude objectives. To be effective here, trainers need to expand their role to train for recovery from error and management of inevitable errors, concentrating on assessing and reinforcing error management strategies.

Error management, at the organizational level, can be defined as having two components, error reduction and error containment. Error reduction consists of measures taken to limit the occurrence of errors, while error containment consists of measures taken to limit adverse consequences. Error management at crew level, is defined as actions taken either to reduce the probability of errors occurring (error avoidance), or to deal with errors committed either by detecting and correcting them before they have operational effect (error trapping), or to contain and reduce the severity those that become consequential (error mitigation). *(Prof. James Reason 1997)*

CRM is not a quick fix solution, just a part (albeit a very important part) of an organizational commitment to error management that includes building and nurturing a safety culture. It has been designed and developed in its 25-year history through collaborative effort between the airline, research and regulatory community. It is an on going, skill based, dynamic development process that provides airlines with unique CRM solutions tailored to their operational demands.

Objectives of CRM training

- To enhance crew (and management) awareness of human factors, which could cause or exacerbate incidents that affect the safe conduct of operations.
- To develop CRM skills and attitudes that when applied appropriately could prevent incidents/accidents whether perpetrated by technical or human failings.
- To use CRM knowledge, skills and attitudes to conduct and manage aircraft operations, integrating these techniques throughout every facet of the organisation culture.
- To apply these skills to integrate commercially efficient aircraft operations with safety.
- To improve the working environment for crews and all those associated with aircraft operations.

Background

Before CRM examiners, check pilots, trainers used to assess flight crew for 'airmanship and captaincy'. Captaincy (or good CRM) is easy to recognise when it is there, just as its absence is obvious, but defining when non-technical skills have reached an adequate standard for safe line operations is still not easy.

Most airlines currently provide CRM training as part of initial training for newly hired pilots. Those training under the US FAA Advanced Qualification Programme (AQP) *(see Chapter 14)* have CRM integrated into their programmes. To effectively address the above issues accurate risk assessment, time assessment, assertiveness and effective workload management must be included in the CRM training provided to new pilots. Training should combine captains and first officers together operating as a crew.

...error management!

During training development, learning objectives for any CRM programme need to be tailored to the CRM principles applicable to 'pilot not flying' (PNF) as well as 'pilot flying' (PF) duties. Using case studies, discussion, line orientated flight training (LOFT), de-briefings will help the new pilot recognize the issues and skills involved.

The FAA have promulgated in depth advisory information on how and why CRM training should be conducted, following on from research into contributory causes of incidents and accidents, 70% of which, contained human factor elements that if corrected, had the potential to beneficially alter the outcome. The European Joint Airworthiness Requirements (JAR's) defines the requirement:

- Management of crew co-operation
- Maintaining a general survey of the aircraft operation by appropriate supervision
- Setting priorities and making decisions in accordance with safety aspects and relevant rules and regulations appropriate to the operational situation, including emergencies
- Exercise good judgement and airmanship
- Understand and apply crew co-ordination and incapacitation procedures
- Communicate effectively with other crew members.

... convince me!

Development

In common with other aspects of flight crew performance, the achievement of high standards of CRM rests on a foundation that consists of several layers. Crew performance will be determined by individuals behaving and operating to a set of skill based standards, requiring them to have knowledge, skill and attitudes that are soundly based on airline and regulatory operating principles. Developing this knowledge, the skills and attitudes depends on trainers behaving and operating to certain standards and have the commensurate knowledge, skills and attitudes themselves.

Training methods need to be focussed on objectives, rather than be activity driven thus avoiding the 'tick in a box' mentality. Some training programmes have, in the past, been constructed and assessed on content, rather than on the basis of objectives. Objectives are defined in terms of measurable outcomes. By focussing effort and investment in training this way we recognise that content is only the means, not the end in itself, to training and education.

Crew standards

It is obviously the task of training departments, and that of individual trainers, to ensure that crews are being trained to the required competency standard. To achieve implementation and standardization it is suggested that the following typical behavioural markers/competency standards be applied:

- When conflicts arise, the crew remains focussed on the problem. Crew members listen actively to ideas and opinions and admit mistakes when wrong – conflict issues are identified and resolved
- Crew members verbalize and acknowledge entries to automated systems
- Cabin crew are included as part of team in briefings – guidelines are established for co-ordination between flight deck and cabin
- Tone of the flight deck is friendly, relaxed, supportive
- Crews adapt to other members personalities and interpersonal differences
- Crews act decisively when situation requires
- Avoid complacency
- Prioritize tasks and manage time for effective accomplishment
- Remain calm and positive under pressure
- Clearly communicates decisions about operation of the flight
- Involves entire crew in the decision making process
- Individuals are encouraged to offer ideas and views and due recognition is given
- Information about problems is clear, accurate and applied with the appropriate degree of urgency
- Potential and actual conflicts are identified and actions promptly taken to deal with them
- Inadequacies in information are identified and alternative sources are sought
- Flight deck crew are encouraged to offer opinions, express concerns and exchange information in an open manner so as to promote trust, mutual support with effective working and decision making
- Breakdowns in communication are recognized, assessed and action taken to improve the immediate and long term situation
- Information is presented to, and sought from all company sources, including cabin crew, maintenance, flight dispatch, ground staff,

operations, crewing etc. in such a manner so as to promote mutual support and effective communication

- Essential activities are maintained whilst collecting information
- Conflicts of opinion are clearly elicited, stated, assessed and resolved via reasoned argument and appropriate evidence
- Effects of any action are constructively reviewed and used to inform the continuous decision making process and flight deck management loop
- Briefings are interactive and emphasize the importance of questions, critique and the offering of information
- Crew members speak up and state their information with appropriate persistence until there is some clear resolution
- Critique is accepted objectively and non-defensively
- The effects of stress and fatigue on performance are recognized.

...when conflicts arise!

The problem areas with non-technical skills

The detractors against CRM non-technical skills (NTS) at airline level offer the following arguments:

- Not enough known about NTS
- The assessment is subjective
- NTS assessment is easily abused
- Cultural differences preclude NTS assessment standardisation.

The way that CRM is currently presented make some of these arguments undoubtedly valid but they can be addressed by:

- Review existing CRM principles
- Identify crew performance problem areas
- Review procedures used by other airlines and regulatory bodies
- Identify possible procedural changes or additions
- Join the many CRM discussion groups both nationally and internationally
- Competency standard for CRM trainers should be established.

...communicate effectively!

Later in this chapter, we investigate the cultural issues of CRM.

Ask yourself:

- What kind of performance do you identify as effective crew performance
- What kind of performance do you identify as poor (or unsafe) crew performance.

For example, specific behaviour can identify either good or bad crew performance if applied to the following:

Crew co-ordination; situational awareness; communication; conflict resolution; decision making; workload performance; risk management; teamwork; leadership; use of SOP's; use of control automation; management of information; passenger handling; flight deck housekeeping; cabin/ground crew liaison; etc.

Identification of CRM procedures should first address the most important crew performance problems. Once training has been implemented, additional CRM procedures can be developed and introduced as an ongoing process, involving the entire organization, constantly seeking ways to improve overall performance.

How CRM procedures work

By directing the training and assessment of CRM skills within crew training programmes, CRM procedures become the focal point, allowing crews to practice specific CRM behaviours in all situations. This helps develop a considered pattern of crew co-ordination, allowing crews to know what to expect.

CRM procedures are an integral part of SOP's and should be integrated within briefings, checklists, emergency and abnormal procedures and be embedded in a whole range of crew activities. During crew assessment, CRM procedures help the instructor/evaluator brief and de-brief the technical performance more objectively. The assessment of a crew's procedural performance is more focussed than the traditional evaluation of general markers. This permits a more accurate understanding of crew performance., hopefully leading to the identification and further development of better targeted training.

Instructor/evaluator/facilitator

The key to combining CRM training and assessment into a well-structured training system lies in sound instructor/evaluator/facilitator training. *(see previous chapter 12)* This requires a sound knowledge and skill in how to brief, administer, assess and de-brief. Assessment skills should be trained in a task specific context, providing the instructor with multiple observations of the range of crew performance that might be encountered.

A thorough crew-training programme should be based on specific behavioural objectives such as those developed under an advanced qualification system (AQP) *(see next chapter 14)*. The programme should be presented in a clear and compelling manner and demonstrate how the application of these procedures improves crew performance.

Crew assessment techniques are an essential part of instructor training and should be based on the collection of reliable data. Trainers can achieve a degree of standardisation and commonality by comparing their rating assessment in relation to their colleagues. The crews, flight operations and the training department are all beneficiaries, in that they are provided with a standard, proceduralised type of CRM, promoting a predicable form of crew co-ordination that is shared and understood by all crew members. It has been suggested that when communication is more predictable it tends to be more reliable and more likely to succeed. Research has shown that predictable patterns of interaction, especially in the areas of crew communication, are associated with better performing flight crews.

CRM development – where do we go from here?

CRM has come along way in its 25-year history, with an astounding and perhaps bewildering array of books, printed matter, articles, forums and development groups available for those who are interested. Regulatory authorities have recognized the importance of human factors and crew resource management and are now mandated subjects for licence issue. Although some of the questions seen recently in one particular examination paper, would have been more appropriate in a book of 'trivial pursuits'. Some of these questions had not been well prepared and thought out and should have no place in mandatory training, as they destroy the credibility of HF/CRM to newly licensed pilots, the very people that matter.

By provoking discussion and argument amongst CRM developers, training departments, trainers, crews and organizations, CRM is evolving day by day, covering an ever widening range of topics. One example of an airline's revamped CRM programme, which they maintain has significantly improved their crews overall knowledge and awareness, was to place emphasis on *skills* based training. Training by this method is 'harder hitting' and more exhausting for the participants as the training department demand that crews arrive at real solutions to the scenarios that are put to them. As an example of the change, they say; in past years we may have been happy with an answer like *'I would be more assertive in this situation'*; now we ask specifically what would they say, when would they say it and how would they say it.

This focuses on issues and items that can be used during normal line operations. A curriculum that blends group dynamics, teambuilding, operational skills, SOP's, and error management and systematically leads individuals, crews and organizations to enhanced skilled behaviour is the strategic aim of CRM. However, what is appropriate for one airline, with its own brand of organizational and national culture may very well be inappropriate for another. CRM if it is to be successful, needs to be tailored and adjusted to suit cultural differences.

...cultural differences!

What is culture?

Simply put, it is the way we do things. It can be defined as the values, beliefs, assumptions and behaviours that we share with others that define us as a group. Culture provides cues and clues on how to behave in normal and unusual situations. Cultural misunderstandings can, and do, occur when people interpret the structure of another culture from their own perspective.

We are all subjected to multiple cultural influences from three main sources:

- Organizational culture
- Professional culture
- National culture.

In aviation, the three cultures can have both positive and negative impact on the outcome of a satisfactory safe flight. The responsibility of organizations is to minimise the negative components of each type of culture whilst emphasizing the positive.

Organizational culture

Senior management plays a crucial role within the organizational culture by influencing the management practices that are noted and followed by its workforce. When employee groups feel they cannot trust management, they will reject with suspicion any new initiatives. It is management's responsibility to provide the leadership, guidance and common vision that is essential in forming and uniting its staff, laying the foundation that ultimately shapes the perception and importance of safety and healthy working practices.

Pilots, in particular, are unique in that they do not have an office where they can feel physically and mentally 'in touch' with the company who employs them. They receive their rosters at home, they report for work, they climb into their aircraft and fly away. On return, they jump into their car and drive home. As a result they can feel isolated and 'left out' with the potential to develop undesirable, suspicious motives towards their employer that is clearly not good. Good management and communication, of course, will ensure that this does not happen but the onus also lies with the trainer and the training department to ensure that during the regular re-current checks this problem of isolation is not overlooked.

...organizational culture!

Professional culture

Pilot professional culture generally shows a great consistency. On the positive side is an overwhelming liking for the job. Pilots are proud of their profession. Being highly motivated they retain their love of the work throughout their careers. On the negative side, they tend to have an unrealistic self-perception of invulnerability to stressors such as fatigue, and personal problems. These are the negative manifestations of the *'Right Stuff'* *(see Chapter 11, Pilot selection)*

The positive components lead to the motivation to master all aspects of the job, to being an effective team member and pride in the profession. On the negative side perceived invulnerability may lead to a disregard for safety measures, operational procedures and teamwork. A safety culture includes a strong commitment to training as well as reinforcing safe practices and establishing open lines of communication between operational personnel and management to threats of safety.

National culture

There has previously been a long held view that the flight deck is (or should be) a culture free zone and one in which pilots of all nationalities accomplish their common task of flying safely from A to B. Certainly, the organizational structure, implementation of SOP's and the aim of the training department should be geared to this end. Research data however suggests that there are substantial differences in the way pilots conduct their work as a function of national culture and that the areas of difference have clear implications for operational safety.[1]

...social understanding!

---

[1] Helmreich & Merritt 1998

This research has established that national and even racial culture has a considerable effect on how individuals perform within the operating environment. North Americans are considered straightforward and liberal, Japanese are quiet and subdued, SE Asians are polite and forgiving, French are vociferous, Swiss are serious, Australians are resilient and vocal, Chinese are obsessed with saving 'face' and the English think they are 'normal' and everyone else is foreign. Of course, these national traits are very much sweeping generalization but contain an element of truth in showing how national cultures differ. What may be good aviation culture in say the USA may not work in other countries that have a culture that is different. Most commercial aeroplanes are produced by western cultures and quite naturally carry with them the western notion of operating them, including crew resource management procedures. People from different cultures allocate different weight and meaning to the same things. Teaching and learning methods here have to be altered and adapted to suit the national culture, if it is to effective. The goal being the same, that is to have a culture that promotes the kind of behaviour that optimizes safety. It should be remembered that CRM works perfectly in compatible flight decks but very badly in the opposite situation.

Cultural values

All people possess certain core beliefs and assumptions of reality that will manifest itself in their behaviour depending on their country of origin. The mind is conditioned culturally at an early age, creating in each of us, social values and constraints that may be very different from those in other parts of the world. When mixing culturally the possibilities for complex and possibly hampered interactions become more obvious. For example, silence can be interpreted in different ways. A silent reaction to a question or request would seem negative to the American, the German, the French, the Italian etc. However, the 'listening cultures' particularly of East Asia find nothing wrong with silence as a response. Indeed, in Thailand, Japan or even Finland, it is considered impolite or inappropriate to force one's opinion on others, being more culturally normal to smile quietly, nod and avoid opinionated argument or discord. These cultures tend to agonize over striking a balance between gaining an advantage and correctness of behaviour. Cultural values dominate the structure, organization and behaviours of these 'listening' cultures more so than is the case in the west, preferring the smooth dispersal of power, the automatic chain of command and the collective nature of decision making.

...giving greater deference!

When comparing the more individualistic 'Western' pilots (e.g. USA, UK, Australia, New Zealand) with the more collectivist, hierarchically structured 'non-western' pilots (e.g. Brazil and other Latin cultures in South America and Europe, and Asian countries like Philippines, Thailand, Malaysia, China etc.) strong differences were observed in attitudes towards command structure and communication flow. During their research[1] found that 'Western' pilots have a flattened command structure that has a less formal distance and greater two-way communication between captain and crew. 'Western' cultures prefer leaders who not only consult with them prior to making decisions, but also prefer direct emotion free communication with the right to logically question anything and anyone. However, the more hierarchically structured cultures differ in that they proceed from an understanding that 'men are not created equal' and more readily accept their place within the natural order. In these cultures, the crew acknowledges the greater power distance by giving greater deference, more formality, more agreement and considerably less challenge to the captain.

In individualistic cultures, information sharing is important but there is an accompanying belief that people have the responsibility to speak up if they are unsure. In collectivist cultures, it is not automatic for people to speak up, therefore greater importance is placed on the captain-initiated, top down communication. This 'power distance', or status distance makes it

---

[1]Helmreich & Merritt 1998

harder for the subordinate to challenge the superior than it is for vice versa. As the distance increases so too does the 'loss of face' threat; *'It is better to keep one's mouth shut and be thought a fool, than to open it and remove all doubt'*

...keep one's mouth shut!

For a multicultural crew, coming from different cultures, the above traits can present a very real problem on the flight deck. The 'Western' captain does not understand why his Latin first officer does not speak up, while the Latin first officer silently awaits his commands; the Asian captain is offended that his 'Western' first officer fails to show the appropriate deference to the captain's authority, and the 'Western' first officer thinks his Asian captain is trying to claim a level of respect far greater than what he deserves. This incongruence may be the cause of resentment and hostility that is likely to impact on crewmember co-operation.

CRM training is particularly vulnerable to cultural influences as its focus is on human interaction. Identifying the cultural type of a flight deck is extremely important, not only for the captain, but for all the crew involved, so that optimum procedures can be adopted and adapted. 'Western' style CRM procedures do not transfer well to non-western cultures without some modification. As multicultural crews become more common with the expansion of global aviation, more attention needs to be paid to understanding cultural differences, as a way of reducing uncertainty on the flight deck.

Without doubt, CRM is a valuable and useful tool in safety enhancement across the whole spectrum of aviation related activities. The challenge for the future is to expand its purpose, its parameters and methodologies.

Performance standards for instructors of CRM

The Royal Aeronautical Society (RaeS) convened a group with support and membership of the UK Civil Aviation Authority (CAA) to work towards the establishment of competence and performance standards for instructors of CRM. After extensive consultation across the spectrum of the industry, in September 1998, they produced an 82-page guide, which is relevant to all trainers. The purpose of this guide is to establish industry standards for trainer performance in relation to CRM. It gives guidance and information to operators, providers of CRM training and CRM instructors, on the necessary standards of competence, as well as how to achieve these standards. The guide describes in detail the knowledge and skills required for competence in the instruction of CRM. Part 1 being a general background and overview. It is recommended reading by everyone who has any reason to be aware of or is interested in instructor competence in this field. Part 2 describes what needs to be known and what needs to be shown, to be considered competent, in each of the three different contexts:

- Simulator; aircraft training; line orientated flight training (LOFT)
- Base and line competency checks
- Ground school/classroom training.

Areas of competence

The guide describes a cascade of increasingly detailed levels of specification of the performance and knowledge requirements for competence in the field of training and development.

- Plan and design training and development
- Deliver training and development
- Review progress and assess achievement
- Continuously improve the effectiveness of training and development.

Each area of competence is sub-divided into units. 12 units have been identified as relevant and applicable to CRM instruction. 6 units are found in 'Deliver Training and Development' and 2 units in each of the other 3 areas. This 'model' guide can be adapted to any kind of training/development in any vocational context. It is therefore an ideal framework for instructors who deliver similar training material in more than one setting i.e. classroom, simulator or aircraft. The competency profile outlined provides the basis for developing a generic competency profile for all flight crew instruction and stimulates a seamless approach to flight crew training in which CRM and technical training are integrated. Refer to *Bibliography* at the end of this book for further information on this guide.

The next chapter looks at some new and different ways of tackling the issues of pilot training: *Total qualification programme.*

The need for CRM training is now well established as an integral part of airline operations.

14 Total qualification programme

Airline training budgets represent a large slice of revenue, and in common with all commercial operations these budgets are forever being squeezed. Most airline philosophy is to buy, or lease time on, a full type specific, national authority approved flight simulator, for both their *in house* conversions courses and recurrent training/checking. They take what they can get from the manufacturer and develop anything else within their own training departments.

World-wide quality and standards vary considerably. National aviation authorities set the minimum standards and are responsible for its policing. Regional operating characteristics obviously influence these standards. However, flying conditions in one part of the world do not necessarily prepare crews for the traffic or weather conditions of other areas. The aim of this chapter is to investigate and suggest that a *total qualification programme* may be an alternative way of acquiring better trained and qualified pilots.

Simulation is the accepted way to train multi-crew commercial pilots. Computer based training (CBT) and various flight training devices (FTD) have their place, especially in type conversion training, but real *hands-on* experience is gained in a full, motion crew cockpit environment simulator. Initial conversion training begins in the classroom using CBT or slide-tape (AV) equipment. First *hands-on* exposure then comes with some form of flight training device (FTD) like a flight system management trainer (FSMT) then the full flight simulator carries the bulk of the training load.

Line orientated flight training (LOFT*)* – Line orientated simulation (LOS)

It has already been stated that nearly 80% of all aircraft accidents and incidents can be attributed to human error. Many of the accidents reports indicate that professional pilots often neglect the human input, by not considering the *softer* issues of flying, i.e. those related to crew interaction, decision making, leadership and resource management. Much has been achieved on improving technical training aspects. One of the main objectives of line oriented flight training (LOFT), is to integrate this technical training and aircraft handling skills with these softer human issues.

LOFT is widely used now by most airlines and substantial resources are allocated for scenario development, perhaps with insufficient attention being paid to instructor training. Poor, unskilled instructors have reduced the training quality of LOFT scenarios, making a significant effect on the outcome of LOFT effectiveness and have been responsible for it being slow to be accepted as a training tool. From the trainee's point of view, the two most important factors affecting LOFT effectiveness are the quality of the scenarios and the quality of the instructor/facilitator's debriefing skills.

...achieving high performance!

Line orientated flight training, should and must, include manoeuvre orientated flight training, (MOFT), if it to be fully effective. This gives the pilot the opportunity to practice and improve aircraft handling skills. With modern automated systems, during routine line operations, pilots get less and less opportunity, to get basic, raw, hands-on flying practice. Manoeuvre orientated flight training, combined with realistic and carefully thought out LOFT scenarios, should be a major feature of any integrated training programme.

Definition

Line oriented flight training (LOFT) or line orientated simulation (LOS), and manoeuvre orientated flight training (MOFT), is a *live* line operating environment flight training programme, with a total crew participation, integrating technical and non-technical training.

Objective

- To practice and develop technical knowledge and abilities to a proficient level
- To improve crew co-ordination, communication and decision making
- To optimize the use of all available resources
- To emphasize the need to operate to standard operating procedures (SOP)
- To recognize your own leadership style and its impact on others.

Training programme

During routine simulator training, pilots will fly typical *real* routes, experiencing some unexpected and *real* events. Knowing that they are able to master these unexpected problems is reassuring and boosts confidence. By using the simulator in this way, crews get practice not only in handling skills, but are able to adapt to changing environmental circumstances, exercising leadership, decision making and judgement. It is rewarding for both the trainee and the trainer to see that the right decisions have been taken. It is instructive to see and realize why perhaps the wrong decisions were taken. Crews are asked to perform a task, as a team, as realistically as possible. Once started with the exercise the initiative is with the crew. The trainer-instructor will act as facilitator i.e. dispatcher, air traffic controller, company representative, cabin staff, or any role demanded by circumstances. During the session he will observe and assess your actions and refrain from intruding as an instructor. He will finally guide the post flight briefing discussion, facilitating the crew to critique themselves. The trainers most active role is most likely to be taken during the initial crew briefing where the basic ground rules of the LOFT would be explained before outlining the scenario.

De-briefings

Trainers who want to expand their knowledge and command of LOFT/LOS should avail themselves of the recently produced, practical guides on how to effectively facilitate debriefings[1]. These guides present specific facilitation tools trainers can use to achieve training and debriefing objectives, suggesting techniques for eliciting active crew participation, in-depth analysis and evaluation including strategies to try when debriefing objectives are not met.

Line oriented flight training (LOFT) provides a method of exposing a crew to a complete flight operation, in such a way that that company procedures, flight procedures, flying techniques and aviation resource management techniques can be observed, analysed and discussed, without the trainer-instructor providing any more assistance than would be available on a real flight. The crew's operation and any shortcomings are reviewed during the de-briefing session. Thoughtful preparation of LOFT scenarios are essential in order to provide a realistic base to fulfil the legislative and training requirements. Good preparation, presentation, review and application are vital, requiring a greater input from the trainer-instructor. The adoption of line oriented flight training highlights this variable human performance factor and is invaluable in alerting crews to this phenomenon.

LOFT facilitation evaluation

The effective trainer/instructor/facilitator/evaluator will lead the flightcrew through a self-critique of their performance and behaviour, including crew analysis of both technical and CRM topics, reviewing positive points as well as those needing improvement. Key learning points are summarized, covering all participants. To guide and evaluate the LOFT the facilitator should:

- State agenda for debriefing and critique
- Ask crew for their appraisal of the flight
- Give his perception of the LOFT, making objective comments focussing on performance

---

[1] LOFT Debriefings: "An Analysis of Instructor Techniques and Crew Participation" – NASA Tech 110442 March 1997 and "Facilitating LOS Debriefings – A Training Manual" – NASA Tech 112192 March 1997

- Show appropriate incidents on videotape for discussion, illustrating behaviours that feature significant points
- Blend technical and CRM feedback during de-briefing
- Ensure that all crew members participate during discussion
- Provide a clear summary of key learning points
- Ask the crew for specific feedback
- Be patient and constructive in probing key areas where improvement is needed
- Be effective in both technical and CRM debriefing.

...competent performance!

The human factor

Pilots are carefully selected, highly trained, routinely checked and yet remain human and subject to human limitations.

Murphy's law says *'if equipment is so designed that it can be operated wrongly, then sooner or later it will be'*

- Human short term memory is rather poor, unreliable, and is adversely affected by stress and fatigue
- Familiarity tends to breed fallibility

- Body rhythm de-synchronisation and sleep disturbance cause reduction in motivation and performance
- Normal personal and emotional problems, as well as operational stresses, have an adverse effect on performance
- Under high workload conditions, fatigue errors are most likely to go undetected, leading to further possible critical difficulties.

Decision making

A continuous process from the time one wakes until the time one sleeps. Not all decisions are unique or critical, most are routine. Flying and monitoring is largely a routine and continuous process, the pitfalls of these routine decisions, is that they can quickly lead to boredom, lack of attention and complacency. Every event does not necessarily need an action, but the first thing to recognize is the need for a decision. This may seem obvious but it is relevant that accident investigators often show this lack of recognition. To be able to make a sound decision it is essential to have as much information as possible, using all available resources.

...leadership!

Leadership

To try to define a good leader or summarize the qualities of one is rather difficult, compounded in flight management terms from ever changing situations, which may require different styles and approaches. However it is important to appreciate that in human resource management terms, a good leader will get more out of people by involvement, basing his decision on their input, encouraging them to think, inform him and judge independently.

Aviation resource management (ARM)

In the mid 1980s, airline training departments and various regulatory authorities, began to look at ways in which pilots might improve their managerial and interpersonal skills on the flight deck. Various terms were coined to describe this practice, flight deck management (FDM), crew management training (CMT), cockpit resource management (CRM), crew resource management (CRM), aviation resource management (ARM) etc. The term *aviation resource management* (ARM) is perhaps the one that most satisfactorily covers all the multitude of resources that exist, to enable a safe, punctual, efficient and comfortable operation. Air traffic control, meteorological services, engineering support, passenger services, load control, company support, catering services, manufacturer support and so on, may all be encompassed under this generic term. Aviation resource management then, describes the ability to use all available means, to achieve the highest attainable level of safety and efficiency within the aviation sphere.

...FDM/CMT/CRM/ARM?

Advanced Qualification Programme (AQP)

In the early part of the last decade, many airline training departments began to rethink their entire flight crew training philosophy. The object was to design a new technology training system that could effectively match technology with training objectives. A thorough analysis of the training needs and available technology was undertaken.

At that time there was growing concern over crew co-ordination and communication, and the effectiveness of existing training methods. Supported and backed by the US Federal Aviation Administration (FAA) some airlines are developing the so-called, *Advanced Qualification Programme*, as an alternative on-going training scheme. The AQP sets proficiency objectives and requires training and evaluation to be conducted in a crew environment. It is a *closed loop* training concept requiring a lot of input, especially from airline training departments, as carefully prepared individual profiles are built up and recorded on each pilot. Training is shifted away from traditional hours, check ride exercises and testing, towards programmes tailored for individual pilots, flying specific type aircraft including evaluation of crew resource management (CRM) with emphasis on line oriented flight training (LOFT) and manoeuvre oriented flight training (MOFT) and human factors (HF).

A majority of the major US airlines now have adopted AQP with its training emphasis on principles and concepts that improve crew performance and flight safety. The comprehensive implementation package includes CRM procedures, training of instructor/evaluator, training of crews, standardized assessment of crew performance and an ongoing, dynamic implementation development process. Design of CRM procedures is based on the airline's perceived specific operational requirements.

FAA Advanced Qualification Programme (AQP)

The US FAA AQP is a voluntary alternative to the traditional requirements for pilot training and checking. Under the AQP, the FAA is authorized to approve significant departures from traditional requirements, subject to justification of an equivalent or better level of safety. The program entails a systematic front-end analysis of training requirements from which explicit proficiency objectives for all facets of pilot training are derived. For pass/fail purposes, pilots must demonstrate proficiency in scenarios that test both technical and crew resource management skills together. Airlines participating in the AQP must design and implement data collection strategies that are diagnostic of cognitive and technical skills.

Features of AQP

- Participation is (at this moment) 'voluntary' – company decision

- An AQP may employ innovative training and qualification concepts
- An AQP entails proficiency based qualification within an approved curriculum
- Each airline develops its own proficiency objectives and when approved by FAA becomes the regulatory requirement
- Are aircraft type specific
- Provide indoctrination, qualification and on going qualification curriculum for every duty position
- Provide training and evaluation in a full flight deck crew environment
- Integrate training and evaluation of CRM and technical skills
- Provide specific training for instructors/evaluators/facilitators
- Collect performance proficiency data on students, instructors, evaluators and facilitators and conduct airline internal analyses of such information for the purpose of curriculum refinement and revalidation
- Integrate the use of advanced flight training equipment
- Airlines are required to submit data to FAA demonstrating that their crews have mastered the appropriate skills
- Provide a systematic way of identifying tasks and sub-tasks involved in a particular phase of flight.

Total Qualification Programme (TQP)

In its present form, it is unlikely that the US FAA AQP will get world wide acceptance, due to its somewhat cumbersome, bureaucratic, input. However, many airline training departments, and regulatory authorities, recognize that there is a need to improve current training methods, which to a large extent, have centred round historical attitudes, and do not wholly prepare crews for current operations. A Total Qualification Programme (TQP), utilizing the basic concepts of the US FAA AQP, of providing a quantitative analysis of pilot progress, appraisal and development is proving to be a workable alternative. A TQP aim is to set proficiency objectives and requires training and evaluation to be conducted in a crew environment. It also allows greater flexibility in the use of new training media and methods.

The key features of a TQP are that it is based on individual pilot proficiency, applied to a single aircraft type, operated by a specific airline. Under this concept, CRM, LOFT and MOFT are an inherent part of the training programme. The captain, first officer and where appropriate, flight engineer and cabin staff, are evaluated and trained at the same time. The rationale for incorporating a TQP is that traditional methods of training are generally considered inefficient, in terms of both economy and resulting pilot proficiency. The major benefit of a TQP is its ability to include and heavily involve the human factor element in the development of the training and qualification.

TQP Curriculum Development

A TQP is founded on the principle that the content of training and checking activities should be directly driven by the content of the operational job. The first step therefore, in developing a TQP is to conduct an aircraft operational specific, job task analysis. Starting with a development of a comprehensive task listing for each duty position, the task listing should cover the full range of conditions and contingencies, including internal to the aircraft, external to the aircraft, normal, abnormal and emergencies that the pilot may be exposed. The task listing would then be analysed to identify the skills, knowledge and abilities necessary for competent performance. CRM skills pertinent to each task should be identified. Additional analysis to support subsequent syllabus development would include such considerations as relative frequency of occurrence in routine operations, operational criticality and success criteria. Some means of identifying and defining applicable performance standards and proficiency objectives will be required not only to prepare individuals and crews for subsequent training in an operational flight deck environment but provide data for approval by the regulatory authority.

Crew resource management (CRM) has become an integral, mandated part of crew training in the United Kingdom. The UK Civil Aviation Authority (CAA) have expressed their concern that delivery of CRM training was proving to be very variable in content and technique and have in conjunction with industry representatives produced a practical guide entitled 'Guide to Performance Standards for Instructors of CRM Training in Commercial Aircraft'[1]. The purpose of this guide is to establish industry standards for trainer performance in relation to CRM. It gives guidance and information to operators, providers of CRM training and CRM instructors, on the necessary standards of competence, as well as how to achieve these standards. *(see Chapter 13)*

---

[1]'Guide to Performance Standards for Instructors of CRM Training in Commercial Aircraft' Published jointly by RAeS, CAA & ATA Sept 1988

Undoubtedly one of the main difficulties in running this type of programme, is how to assess the aviation resource management and human factor aspects in order to achieve some consistency. Trainer-instructor training is a vital element if TQP is to succeed. It does require a lot of input and application. Airline training departments need to look again at the way some of them use their simulators and other training devices, to devise ways in which they can be better utilized. However, it is appreciated that as the degree of line orientated evaluation grows, there will be a corresponding increase in workload for trainers and evaluators. Success of such a programme can only be measured by the results it produces in the years to come. The pioneering work being carried out by some training departments, will hopefully result in broader and deeper understanding, which will benefit everyone concerned with safety and economics of the airline industry.

GOLD

BRONZE

SILVER

...quality control!

Summary

To some extent, many airlines already operate a mini type of TQP for re-current training and checking. This is still somewhat hampered by some regulatory authorities (UK CAA for example) insisting on their so-called TQP, the bi-annual proficiency check and annual instrument rating check, which historically use traditional training assessment methods that are created somewhat artificially. Being event led they tend to lead to the 'tick in a box' syndrome. Indications are however, that they and many other authorities are widening their brief and are recognizing the value of the total qualification programme (TQP) as part of the legal requirement for maintaining standards.

Creating an *in-house* TQP is certainly a formidable task for any training department and generates a lot of input, as each pilot will have his own personal programme requirement. It will not just be a tick in a box. LOFT/MOFT/ARM/HF are an integral part of each training session. Assessing these *softer* issues does require some skill, training, practice, patience and experience on behalf of training staff, if it is to work effectively.

...patience!

Some countries still consider the qualification of the individual pilot is more important than the crew as a whole. Whereas some emerging third world and newly democratized countries buy modern western built aeroplanes, but do not have the expertise to operate them safely. The rapid growth of airline route structures and fleets, resulting in crews, with relatively low experience and training levels, being promoted to fly complex contemporary aircraft has not improved accident statistics.

In reviewing this it is acknowledged that the design of aircraft and operating procedures are predominately biased towards Anglo-Western culture. Western culture tends to be individualistic whereas many (but not all!) Asian and Latin American are considered collectivist. In the former, individualism is highly regarded, whereas collectivism reflects inter-dependence amongst a group.

Research indicates that most if not all cultures, given proper selection and training, are capable of maintaining good co-ordination, technical proficiency and discipline. However, when crew co-ordination breaks down, all cultures are vulnerable to error, either by commission or omission, whether it is an autocratic individualistic captain, or supporting crew reluctant to question a commander's authority. Recent serious accidents have highlighted the need for better crew resource management (CRM) training and awareness.

As we enter the new millennium, commercial aviation is facing a big challenge. The successful execution of a total qualification programme (TQP) is critically dependent on the establishment of a systematic development and quality control process:

- Establishment of a comprehensive, web-accessible data base containing detailed task analyses of every make, model, series and variant of aircraft
- Standardized implementation of software tools for curriculum design so that required tasks, skills and knowledge may be systematically audited and evaluated
- Advancement of 'state of the art' training, evaluation and task analysis
- Uniform establishment of data driven quality control tools for use by the airline training departments and regulatory authorities
- Establishment of a continuous system of curriculum refinement and improvement
- Creation of full flight simulator scenarios that precisely target technical and crew resource management (CRM) skills
- Effective integration of CRM and technical skills training evaluation throughout the entire curricula
- Uniform implementation of a system that quantifies and maintains a high level of standardization and quality amongst instructors/evaluators
- Integration of subjective and objective assessment of pilot proficiency
- Maximize efficiency and cost-effectiveness of training methods
- Monitor the reliability and validity of pilot proficiency measurement.

Of course, the most sought after benefit of a total qualification programme (TQP) is that there should be better trained and better qualified crews, giving a much needed boost to overall flight safety. It is expected that within the next five years all major US airlines will have adopted this style of training.

The next chapter draws attention to some of the problems encountered with current training practices.

Airline Flight Crew Centre.

15 Tomorrow's training today

In 1992, over six hundred members of the British Airline Pilots' Association (BALPA) took part in a survey to find out what *they* thought about their type conversion training. The majority felt that aircraft type conversion training left them with insufficient knowledge of some aircraft systems. This lack of knowledge was cited by the pilots, as being the greatest perceived threat, to doing their job well. Criticisms of ground school training could be summarized as being too compressed and too 'need to know'. A combination of training methods was felt appropriate to learn about new high technology aircraft and systems.

A survey in 1997 examined 1,718 commercial airline pilots' evaluation of the training they received for use of aircraft automation and their attitudes toward the use of automation. One quarter of pilots felt that initial training did not adequately prepare them for operating their aircraft. Overall, these results allow identification of potential threats to safety with the crew automation interface.[1]

Yet another survey studied data from 2,600 line observations from 4 major US airlines. Results indicate there are performance problems specific to advanced technology aircraft. 'A notable percentage of crews showed substandard performance on the line. There is increasing evidence that crew co-ordination and communication in advanced technology aircraft is different, and can be inferior from that found in traditional aircraft cockpits. While crews may be well trained in overall technical proficiency there was still a greater variability with respect to overall team effectiveness'[2]. An on-going examination of data from more than 10,000 pilots from five cultures yielded highly significant national differences in both general attitudes regarding the conduct of flight and the use of automation. Pilots from different cultures were found to hold widely

---

[1] Source: 'Aircrews' Evaluations of Flight Deck Automation Training and Use' P. Sherman Univ. of Texas July 1997
[2] Source: 'Flight Crew Performance in Standard and Advanced Technology Aircraft' W. Hines Univ. of Texas April 1997

divergent attitudes. In high power distance (PD) cultures *(see Chapter 13 Culture & CRM)*, superior-subordinate inequality is accepted by both superiors and subordinates, with organizational command structures being hierarchical and rigid. These cultures are more likely to accept authority, for example by perceiving the autoflight system as a highly 'competent authority', they prefer to rely on its expertise. In low PD cultures, subordinates do not usually favour close supervision and are more likely to question a superior or take action in ambiguous circumstances. Being more used to an interactive style these cultures feel more comfortable taking an active and controlling stance with automation. When comparing the two, mistrust and reluctance to utilize automation's expertise may be as detrimental as blind over-reliance on automation. A philosophy of automation use situated between these two extremes should be the aim.[1]

Automation

Automation is not error free it introduces different kinds of errors, some of which are not immediately apparent. Examples are incorrectly loaded runways, standard instrument departures/arrivals (SIDs/STARs), winds or co-ordinates. Automatic fuel monitoring for the Boeing 747-400, compared with the B747-200 flight engineer's almost full time job, means that often there is too little for the pilots to do, resulting in a low arousal state. Another poorly designed and programmed flight management system (FMS) calls for 245 button pushes to programme a short 35 minute route, plus another 45 button pushes for the other pilot to cross check the inputs. This type of situation is both time consuming and is a cause of great annoyance. A case of where *automation* really has gone to far, and yet not far enough, and is no help to the pilot.

System errors

There is growing concern that automated systems can work or fail in ways that are both unanticipated and untrained. This difficulty in detecting system errors requires the crew to cross-check primary flight and navigation displays to ensure proper performance of the automated systems. After more than a decade of experience with these advanced systems the promise of improved flight safety is still largely unfulfilled.

---

[1] Source: 'Attitudes Towards Automation – The Effects of National Culture' P. Sherman & R. Helmreich Univ. of Texas 1998

A review* of 324 accidents and 470 incidents found that 23% and 49% respectively, resulted from errors in flight deck management. The human pilot depends more and more on automated aids to assist in control and monitoring of the aircraft and its subsystems. These automated systems are programmed to perform certain goal orientated functions. With both human and automation, performing flight deck functions there is a potential for conflicting goals.

The A320 crashes in Strasbourg,[1] Habsheim,[2] and Bangalore,[3] are classic examples of the pilots failing to interface with automation at the higher levels of technology. More recently we have seen the A300 accidents in Nagoya° and Chiang Kai Shek^ and the B757˘ Cali accident. It is apparent that in many cases we are not applying some of the human interface lessons learnt in the real world. All too often, the pilot is left out of the loop. We must not underestimate the training required for new technology. At Strasbourg, Habsheim, Bangalore, Nagoya, Chiang Kai Shek and Bogota, the automation worked as it was designed to, but the interaction between the machine and the human failed.

Flight control computers (FCCs), giving flight envelope protection, and automatic rudder inputs, can lead pilots into a false sense of security. More training in mode awareness, and human centred automation, are required to keep track on what the pilot has asked the aircraft to do. The human operator must be kept informed, and retain the authority to command and control the situation.

Pilots generally are very positive about the effectiveness and safety of modern EFIS equipped aircraft. EFIS flexibility and capability is of enormous benefit. Very few pilots would prefer to go back to the old displays of electromechanical instruments and manually tuned navigational aids. The importance of scan rates during instrument flying has always been stressed, no less so in an EFIS cockpit. Emphasis however moves away from the old 'T' panel scan, where primary flight information is obtained from six or more basic instruments, to a scan of the primary flight display (PFD) and navigation display (ND). Here all the information is usually presented on just two screens (CRTs). The basic means of

---

[1] Air Inter A320 Strasbourg - July '92

[2] Air France A320 Habsheim - June '88

[3] Indian Airlines A320 Bangalore – Feb '90

° China Airlines A300 Nagoya, Japan - April '94

^ China Airlines A300 Chiang Kai Shek - Feb '98

˘ American Airlines B757 Cali – Dec '95

Source: *NASA Ames Study NASG 2-875 Jun '97

determining automatic flight status is usually by reference to a flight mode annunciator (FMA) on the PFD. The secondary means is by checking the illuminated switches on the various selector panels. Understanding of, and correct use of, FMA information is crucial to safe operation and requires special emphasis during training. More often than not malfunctions and reversions are user induced, or flight status modes have been incorrectly identified.

Flexibility of use will ensure that the pilot frequently allows several different combinations of modes, to achieve a desired result. He will then become to appreciate the advantages and limitations of the various combinations. Trainers should be particularly alert to the possibility of EFIS/FMS distractions that can detract from general flight management awareness.

Pilots transiting to EFIS equipment will find the concepts of lateral and vertical navigation, via the FMS, new. The principles involved will require detailed attention. Whereas pilots transiting, say from the B737-300/400/500 to the B757 or B767 will need to have the differences between, for example, the Sperry and the Lear Seigler FMS highlighted.

...a false sense of security!

Due to the shortcomings of most current FMS, even though approved for primary navigation, basic airmanship demands cross checking of FMS position against derived position, using raw data, from whatever aids are available. In-flight philosophy should be for FMS programming to be done primarily by the support pilot, then cross checked by the operating pilot, before executing any changes. During late descent, approach and landing phase, FMS entries should be restricted, as attention can be better directed to monitoring the aircraft and keeping a good look out. General flight management requires the flight to be operated within the limitations set down, either in the company operating manual, or by the national regulatory authority. To operate within these requirements the pilot needs to have a good understanding of all aspects of these limitations and their practical use.

Trainers should emphasize the need to stay ahead of the aircraft especially when approaching critical flight regimes. A timely request for ATC clearance, careful descent planning, obtaining weather (ATIS) information early during quiet periods, are all examples of good flight management. Acclimatisation and orientation within the cockpit environment can form a very strong link in the chain, enhancing flight safety. Lack of knowledge of, and the location of, the various controls and emergency equipment will give rise to an uncomfortable environment, degrading performance.

In spite of what we have inferred in the proceeding paragraphs, automation has made flying safer, more economic and more reliable, however it must be there to help humans, not to hinder or replace them. Humans are better decision makers than computers. The problem for the training pilot is in getting to grips with this relatively new man-machine interface adapting and adopting his training methods to suit.

...EFIS distractions!

Electronic Flight Information Systems (EFIS)

EFIS, as well as being a combined flight and engine instrument display, is also a primary monitor of selections and automatic functions in flight. The flight management system and its display are an integral part of this man-machine interface. It is now standard fit on all new major transport aircraft. The production and maintenance skills required for electromechanical instruments have made them redundant. The effectiveness and flexibility that EFIS brings to a modern flight deck is without question. However the flight controls and displays are capable of enormous power, and it is this, that is the cause of concern, when considering some recent attributable accident reports.

Some of the present airline and manufacturer training courses give an insufficient overview from the start. Trainees then have difficulty in building up the detail, to see the whole. Flight management system messages can be perplexing and impracticable, encouraging cloistered vision in solving small problems. Aircraft management then suffers. Conversion training courses need to address this problem by stressing basic aircraft first, then a gradual integration of exposure to automatics. The flight management system, as on any computer, needs the user to be proficient in all the formats and inputs. Trainers should be particularly aware of, and point out to their trainees, the compelling, distracting nature of modern EFIS displays. The pilot needs to keep the essentials in mind, of attitude, performance and navigation. Automatics can, and do, go wrong, there is as yet, no such thing as a fully protected aircraft. An over focus on a perceived primary source of information will cause other important data sources to be monitored less thoroughly.

Total qualification programme (TQP) simulator training *(see Chapter 14)* gives the opportunity for pilots to be refreshed regularly on full system capabilities and to identify aircraft management shortcomings.

Interestingly, the manufacturers of aircraft and navigation systems do not, as yet, guarantee the accuracy of data in the navigation data base, neither is the data checked in the flight management system during certification. Actual deficiencies vary in magnitude, from those that cause an increase in pilot workload through confusion and overload, to those that result in errors in navigation. Even two linked FMS navigation databases will guide the aircraft on slightly different courses. The pilot will only recognize the error if he cross checks the aircraft's position, using raw data from individual navigational aids. Advances into future air traffic management systems will not only require better accuracy of navigation databases, but different emphasis on standard operating procedures.

A pilot, perhaps more than ever, needs to keep the essentials in mind, position, attitude, performance, navigation and an overall situational awareness. Airline training departments, in many cases, need to look again at the effectiveness of their training programmes. Flying today is much more complicated; some of the tasks that have to be taught were not heard of ten years ago.

...automatics can go wrong!

In the mid 1980s the University of North Dakota's Aerospace Department carried out a fundamental study of the skills and knowledge needed to become a competent professional pilot and crew member in a *glass cockpit* of today's modern jet airliners. During the two years of research, they identified about 30,000 component tasks. The written definition of these comprised 1,000 pages of single spaced material!

Today's pilots, who only complete their ab-initio training to national commercial pilot's licence and instrument rating standard, on light piston twins, are hardly equipped to cope with the complexity, speed, size and inertia of modern jet transports. It should be appreciated that often these pilots join their first airline in a completely different kind of operational environment from that which their instructors and training captains originally trained. A research project in the US determined that it takes at least 600 hours of flight operational time for crews to transition from novice to master on today's automated aircraft.

There is still no industry standard for automatic flight and flight management systems. Different manufactures produce different designs, which all look similar but have subtle differences. FMS differences are difficult to assess, as pilots tend to like what they have mastered first. It would be helpful for example, if manufacturers moved to a common EF1S terminology. Formats of EFIS, FMS charts and company documents are still, in many cases incompatible.

The undesirable EFIS display

Occasionally EFIS/FMS system designers do produce questionable layouts and presentations, failing to take account of human susceptibility to err when information is not presented in the most acceptable form. Following the Kegworth[1] B737/400 accident in 1989, a contributory factor cited by the UK Aircraft Accident Investigation Branch (AAIB), was the aircraft's combination of layout and dial presentation. Studies were done which showed that the presentation on this aircraft was the most inefficient of the various permutations possible, resulting in 60% more reading errors than the best, and taking 25% longer to read.

Some the latest FMS equipment, although giving superb three-dimensional precision and presentation, could hardly be described as user friendly. Some of them create difficulties in overall management, with them being hard to change functions or access new information. A system where incorrect way-point entry removes the previous one, for example, raises the workload. The pilot then has to revert to a rough heading from memory and work things out. The UK Civil Aviation Authority (CAA) recently turned down one FMS on being below expected usability. Some of the current cockpit EFIS displays are engineers designing what engineers believe pilots need. Hitherto, the influence of pilots on cockpit ergonomics has been less than it should be. It is undoubtedly a controversial area, too much complexity induces confusion, too much simplicity induces complacency. Both cause accidents. However, pilots should not have to rely on training and procedures to overcome unnecessary design flaws.

EFIS displays can seem so complicated that they are seductive and risk distracting the crew from basic airmanship. An over focus on a perceived primary source of information will cause other important data sources to be monitored less thoroughly. Only with proper training can this be smoothed out.

---

[1]Source: AAIB report British Midland B737/400 – Jan '89

...good decisions are based on good information!

Advanced cockpit design

In the last decade, we have seen fundamental and innovative changes to flight deck design. The Boeing 747 original design, for example, goes back to the mid-sixties and yet in its latest form the B747/400 the cockpit bears little resemblance to its predecessor. This is a digital avionics aeroplane. Airbus Industries family have revolutionary side stick control as well of course computerised *fly by wire* controls. This system is universally liked except for two questionable design features universally disliked by most pilots. *(see Chapter 4 on the importance of perception and stimulus.)* The control side sticks move independently, the only way one pilot knows what the other is doing is by observing what the aircraft is already doing. The aircraft reaction is the sum total of the two sticks. Additionally the power levers, whilst being managed by the FMS, in cruise, do not move when the engine power is changed. Both these features have already caused incidents, (tail strike on rotation when both pilots put in input to avoid a flock of birds and several cases where pilots have manually moved the power levers incorrectly). This removal of tactile information is another example of poor cockpit design and arguably makes the system less effective and therefore less safe.

There are still, yet, no national or international standards for flight decks, or its displays, there are no benchmarks for clarity, user friendliness or for cockpit layout and design. Pilots will put up with unclear instruments and poor layout because they are required to be flexible and mould themselves to requirements, however unreasonable. It is particularly characteristic for pilots to set themselves to overcome flying challenges. For that reason, the manufactures and the airline establishments are more likely to deal with a cockpit imperfection by modifying pilot training rather than change the design. Although having said that, things are changing. The manufacturers of the new generation of aircraft now being produced are going to great lengths to address this problem, by involving airline customers and pilots, at the design stage.

Accident prevention

There is a school of thought that pilot training is becoming too concerned with managing systems, and too divorced from instilling basic flying skills. Current training, on the latest generation of airliners, conditions the pilot to leave the basic handling to the aircraft's automated systems. This can only lead to declining standards of physical piloting skills and overall judgement.

Analysis of accidents, during the past ten years, reveals that only eight per cent were caused by a single factor. In all other cases, there was more than one prevention strategy available. Prevention strategies and actions are carried out by pilots on every flight, and the flightcrew is usually the last part of the system that could prevent an accident. The implications of possible declining standards of piloting skills, by increased automation, need some special consideration.

Avionics react quickly and correctly, often more so than humans can. Many pilots using EFIS equipped aircraft are concerned about losing handling skills, and the inevitable question arises, what happens if the automated flight control systems and their multiple backups fail? Of course a pilot must be trained, and is trained, to handle situations and unexpected problems. However there has been a lack of training on how to detect and handle an unexpected situation, when one gets the feeling that, something is wrong, *but what!*

The psychological factor

Psychology is now recognized as an important element in airline training departments. The gradual awakening, to the fact, that the human element is as important, as technical skills, in running a safe operation. Training departments need to see the development of human psychology, as an essential part of their task, to do it professionally and stop relegating it to a left over from technical training, as is now, so often the case.

Although a great deal of effort goes into the design of modern flightdecks, it is not possible to foresee and prevent all potential problems. Ergonomic failures still contribute significantly to incidents and accidents. The increasing reliance on computer software makes it even more important that airline crews understand how equipment, software design, and human limitations can interact in potentially dangerous ways. A fatal accident to an A330, albeit on a test flight in Toulouse, highlights some of these problems. The commission of enquiry within the DGA[1], the French military body which investigates test flight accidents, has stated that a combination of several factors, no one of which, in isolation, would have caused the crash. Ancillary factors, included in the report, stated that lack of autopilot mode visual indication, obscured at high pitch attitudes, crew confidence in how they expected the aircraft to react, the delayed reaction of the test engineer to assess the developing parameters, the captain's slowness in reacting to the developing abnormal situation, as well as numerous small errors contributed to this accident.

As aircraft become more reliable but also more complex, the role of the pilot is changing. The modern pilot *is* more divorced from both the environment and feel of the machine. The pilot of today has more pressures now than those purely derived from being on the flight deck.

...pressures!

Crew resource management (CRM) training is now a major feature of all modern airline training departments. Indeed, it is a mandatory subject for licence issue by an ever increasing number of regulatory authorities. Basic aviation physiology, psychology, stress, fatigue, social psychology and ergonomics on the flight deck are typical subjects covered. *(see Chapter 13)*

Attitudes and personality

Pilots generally are intelligent, receptive, have high integrity and are used to receiving information. Airline training departments wishing to run successful crew resource management (**CRM**) courses need to ensure that their courses are aimed at enlightening and possibly changing attitudes.

An attitude is a variable, inserted between reality, and the way in which we respond to, or experience reality. As human beings very few of us scrutinize our own attitudes. If a person's feelings are strong and opposite, he will block his perception of the information, deny its relevance, or even claim the information is wrong, stupid or a lie. A lot of airline pilots have a negative attitude towards psychology. A negative, wrong, or unsuitable attitude can create personal stress, illness and unsafe conditions on the flight deck.

Problems with personality can arise when people are required to work in together in confined areas. The stereotyped captain is still often portrayed as being arrogant, self-opinionated, domineering and intransigent, whereas his stereotyped first officer would be seen as being aggressive, impulsive and uncooperative. Understanding personality profiles, both of yourself, and your colleagues, is an important step forward, both in human factor training courses and in smoothing everyday working practices.

...with a broader vision!

Things *are* changing, yesterday's traditional crusty, rigid chief pilots and managers, are being replaced by a new breed of people, hopefully with a broader vision, who are more open to change and new ideas.

There is now a gradual awakening to the fact that the human element is as important as technical skills in running a safe operation. It is understood that you cannot manage people without understanding them, anymore than you can manage a complex airliner in flight, without some technical skill and practice. A good grasp of psychology can enable you to bring about changes, both in yourself, and your work environment. As an airline pilot, unless you are willing to understand yourself and give some priority to your own growth, you will be ill equipped to understand others and help them develop. Get to know yourself. Understanding your own assumptions, motivation, needs, and fears, is the first step towards making positive improvements.

A deadly combination is a bad work environment and the prevalence of personal *hot potatoes*. Good safe operation results not just from joint technical skills but from how people function as a team.

- Familiarize yourself with those psychological theories and techniques that can be useful to you
- Examine your own psychological make up and provide opportunities for your trainees to do likewise
- Recognize how you are currently helping or hindering each other from operating safely and efficiently
- Identify the changes in attitudes and behaviour that you want to make and take active steps to put all of this in motion
- Set the tone for a new style of problem solving in the future.

...personal hot potato!

A lot has already been said about decision making and communication on the flight deck. The quickest way to hamper or destroy decision making and communication is to create the unexpected. Airline training departments and its training staff's main function, to cope with ever increasing complexity of modern aircraft, is to give better training in how to minimise the possibility of external causes reducing the pilot's cognitive processes, when under time stress. Very high stress levels, can, and do impair the ability of discriminating between the relevant and the irrelevant. Modern training methods *should* produce a tailored pilot, showing less variation in performance.

Sex differences

Understanding sex differences in performance evaluation, has been the subject of very little study or research, although it is still the butt of many sexist jokes. Earlier studies that were undertaken often showed that the performance of equally qualified women were rated as lower than that of men. However more recent research of *real* people has pointed towards little difference in the rated performance of the sexes, and indeed shows a slightly superior rating for women.[1]

A case is made that piloting aircraft requires similar skills to managing other complex enterprises and that similar profiles predict success in each. The focus of this study was to attempt to discover how personality relates to evaluation by individuals, at all levels, and whether certain personality characteristics are differentially valued in both male and female managers. In this research, in actual organizations, women did not show a rating deficit. On the contrary, their ratings were significantly higher than those of male managers. Strong confirmation of the relationship between personality and assessed quality of leadership was discovered using the cluster* analysis to establish constellations of personality.[2] Comparisons among the five clusters on each personality dimension found statistically significant superiority, for the highest rated managers, on

---

[1]Source: 'Managerial Leadership Assessment, Personality Correlates of and Sex Difference in Ratings' Robert H. Gibson and John Wilheim - Department of Psychology, University of Texas at Austin –1989
[2]Source: Validating Personality Constructs for Pilot Selection - Status Report on the NASA/UT Project - Robert L.Helmreich and John A. Wilhelm - Department of Psychology, The University of Texas at Austin - March 1989

attributes and motives typically associated with achievements. A parallel analysis found that females had higher scores than males on *positive* attributes and lower scores than males on negative attributes.

So how different are the sexes?

Females certainly approach problems in different ways to males. Simplistically, women consider options, men look for solutions, men value status and competition, while women perhaps encourage co-operation. Men tend to ration and control power, while women share. Men want respect, women seek rapport. However, women are often ignored or dismissed through their less insistent presence, having less pervasive body language and quieter voices. Men tend to talk about power, their cars and personal achievements, whereas women tend more to concentrate on their relationships with staff and personal matters.

So typically, the sexes do exhibit different characteristics. So what! How does this make any difference to airline operations? Well data suggests that although female pilots are very much in a minority, they do have to possess stronger personal characteristics than men, to achieve comparable positions. However when considering personality, their positive attributes are often far more effective than men, most useful in the field of human factors and aviation resource management.

Automatic flight v manual flight

One of the many problems facing training departments is the increasing sophistication of navigational, approach and landing systems. Crews are routinely expected to land their aircraft in weather conditions that a few years ago would have resulted in a diversion or delay. The right balance between maintaining manual piloting skills, to cope with the occasions when automation is not available, and fully automated approaches has to be considered. A further problem is that as long haul aircraft fly further and longer than they used to, they consequently make fewer take-offs and landings, giving pilots less opportunity to practice their skills. An industry problem!

Automatic flight capability reduces the time available for a pilot to manually practice basic skills. Every pilot should therefore endeavour to set personal exercises, aiming to maintain and improve his own standard. This can be achieved by regular manual flying to keep current. However, it is important that targets are set:

- Fly speeds accurately
- Fly nominated bank angles precisely
- Fly manoeuvres at a pre-planned rate
- Descend-climb at a pre-set rate while decelerating-accelerating
- Where an altitude requirement is set capture smoothly and hold it whilst compensating for speed, power changes.

... what if... ?

In flight cockpit scans are an integral part of flight management, but they also limit one's perception by stereotyping actions. Trainees should be encouraged, for example, after completing a cruise-climb check, to ask themselves, *what if...?* rehearsing the various abnormal procedures.

Computer based instruction, and computer based training, have the advantage of giving the pilot *hands-on* practice, which is vital for operating computer and electronic systems. It also enables pilots to try out various scenarios, which clarifies the operation, and demonstrates the need for standard operating procedures. However great value is still placed on having experienced human teachers as well. We must not lose sight that aviation, in all its many guises, is a human endeavour.

Licensing rules

Historically individual countries have controlled and issued their own crew licences. Beginning in 1999, European pilots from the 27 member states of the Joint Aviation Authorities (JAA) have agreed to operate a common licence. This licence will be issued under the Joint Aviation Requirements -

Flight Crew Licensing (JAR-FCL) framework and will be recognised throughout the participating JAA countries. These countries have agreed on a common training and testing requirement for pilots. Placing emphasis on the training structures to be used, the system is closely based on, what are regarded as, the two toughest training and licensing regimes in Europe, namely Germany and the UK.

Student pilot training

Under the JAA framework more emphasis is placed on the concept of student pilot in command, through which the trainee assumes more left-seat responsibilities, this requires a greater instructor input than before. Designated chief flying officers are based at each approved location, being personally responsible for ensuring JAR-FCL compliance and issuing the course completion certificate. National aviation authorities will routinely audit this structure on behalf of JAA.

A new facet of JAR-FCL is that a new pilot's first type rating for a multi-crew aircraft has to include a multi-crew co-ordination course consisting of 15 hours dedicated simulator flying as well as 25 hours of integrated synthetic flight in an approved flight navigation procedure trainer. The synthetic and simulator elements are integrated with actual flying training.

Discord still exists, in flight training philosophy, between Europeans and Americans. The Europeans have always set great store by in-depth and concentrated theoretical and practical training, whereas the US emphasis has tended to be on shorter training supplemented by 'hours building'. New pilots may be attracted to the relatively low cost of flight training and acquiring a US licence, but will, if they require to convert it to a European one, have to usually undergo extensive conversion training and testing.

Food for thought

From 1980 to 1996 there were 287 approach and landing accidents (ALAs) world-wide resulting in 7,185 fatalities involving Western built transport aircraft. The average ALA rate is 14.8 fatal accidents per year. If this trend continues then 23 fatal accidents can be expected each year (one every two weeks!) by 2010.[1]

---

[1]Source: Flight Safety Foundation – approach & landing accident reduction

Controlled flight into terrain (CFIT), despite a world-wide campaign to reduce it, increased in 1998 for the second year running, confirming a reversal of a previously favoured trend. 13 loss of control (LOC) accidents in the same year have caused many agencies to voice their concern. Discrepancies in the levels of safety achieved in different regions of the world should re-inforce the policy of attacking these areas of safety weaknesses.

Studies by both the International Civil Aviation Organisation (ICAO) and the International Air Transport Authority (IATA) have both recorded a big rise in crew errors resulting from insufficient pilot knowledge of aircraft systems and procedures. These deficiencies, classified as H3 errors by ICAO, have been seen to rise fourfold in aircrew flying automated flight decks. The cause suggested by these industry studies focuses on the combination of increased aircraft complexity and *the failure to adapt training appropriately to the needs of pilots flying modern aircraft.*

Concorde 002 first flew on April 9 1969 and has achieved nearly a million hours logged to date (600,000 of them at supersonic speed). The A320 flight deck shows how things have changed in the past thirty years.

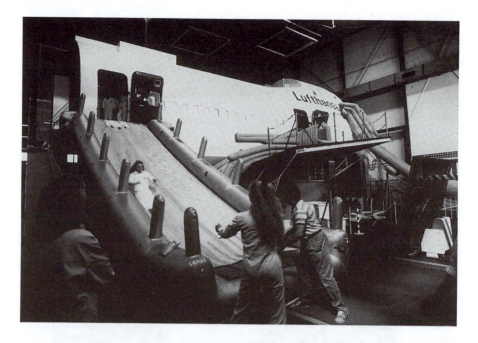

An often neglected part of many airline crew training departments. Practised emergency drills and procedures enhance the crew's overall operation in terms of resource management, leadership, team building and professionalism.

16 Where do we go from here?

Over fifty years ago, on Wednesday 27 July 1949 the world's first jet airliner, the de Havilland (DH) Comet made its first flight from Hatfield airfield near London. When that Comet flew, less than 40 million people were travelling by air each year. Annual passenger traffic is now approaching 1.5 billion and growth continues at over 5% annually. Undoubtedly the development of the jet airliner has made this possible. Market research indicates that over the next twenty years some 17,000 commercial aircraft will be required by the world's airlines, driven by this projected traffic growth and the retirement of existing fleets. Of these about 5,000 will be wide-bodied aircraft, 4,000 single aisle aircraft with a seat capacity of more than 120 seats, 3,000 regional jets in the 50-120 seat capacity and finally around 5,000 turboprops.

Airlines are demanding ever more efficiencies and lower costs from their equipment and staff. Pilots can expect to see more demands for higher productivity, changes to working practices and an increase in all the related problems that these measures will induce.

The new breed

Predictions show that there is a potential need for up to 500 aircraft of more than 500 seat capacity. Increased traffic on long distance routes, coupled with progressively congested airports and skies, will be the driving force behind the next generation of very large commercial transports (VLCT).

The successors to the current Airbus A330/340s, Boeing 747-400 and MDC MD-11s, will seat between 350 (current B747) and up to 1,000 passengers. Airbus has a design concept, A3XX, the stretched version weighing in at around 500 tonnes, and seating around 650 three class passengers, or 1,000 single class. As part of their B747-X programme Boeing are studying designs for a B747 stretch, featuring a new double deck design and folding wings, to fit existing airport gates.

No son and heir yet in sight for Concorde, which first flew in 1969. Few hold less than anything in awe the technical ground the aeroplane broke in an era when computers were virtually unheard of. However, Aerospatial, British Aerospace, and Deutsche Aerospace have formally launched a joint supersonic transport (SST) research programme. This aircraft is intended to have a range of 10,000km (5,400nm), 250 passengers, three class capacity and a cruise speed of mach 2. In the US, NASA has a current $150m development budget for similar supersonic research. Massive development costs are, however, likely to see any *Son of Concorde* emerge as a joint global project.

The real future for long distance travel has to be in shortening journey times, as anyone who has endured a long cramped flight will no doubt agree. Although, still a long way off, there are studies ongoing for a hypersonic transport that would stay in the atmosphere and cruise at mach 5/6.

~~slave ship 1750~~
air travel 2050

...a long and cramped flight!

Cockpit of the future

Pilot's who fly tomorrow's aeroplanes will be the beneficiaries of tremendous advances in technology, some of which are already with us. The way pilots communicate and navigate will be vastly improved, with ATC messages being sent by digital datalink rather than by voice, navigation will be based more on satellites rather than land based navigational aids and traffic surveillance will be achieved by networking tactical information between all aircraft. Displaying large liquid crystal (LCD) 3 dimensional displays, cursor control devices, voice recognition systems and head up displays (HUD) the future cockpit will be vastly improved.

New subsonic aircraft are most likely to evolve round current types already in service. Featuring lighter, more advanced structures, leaner, quieter engines, more automation and high technology navigational systems.

Airlines lose an estimated staggering $3.5 to $5 billion US dollars each year as a result of inefficiencies in today's ATC environment. Departure and arrival slot times, stacked arrivals, gate holds and delays together with flight at inefficient altitudes and vectoring on circuitous airway routes add to the cost of the business. By incorporating sophisticated data linked avionics in a 'free flight' environment, pilots will be released from the rigid discipline of being spaced in nose to tail time blocks, along less than optimum routes and altitudes.

Already under development, using technology derived from defence applications, are various enhanced situational awareness systems (ESAS). Information from ESAS is intended to be used in conjunction with data from position inputs from global positioning systems (GPS), satellite navigation or other sources. The database of an on board electronic library system (ELS) will enable aircraft to position themselves with greater accuracy relative to a runway. Future autonomous aircraft will be able to operate safely, even into the most isolated and primitive airfields with the minimum ground based aids. ESAS combined with head up displays (HUD) could well form the basis of the next generation of autoland systems. The so-called *hybrid* approach might consist of fail passive autoland system, with a fail passive HUD as monitor and back up. A suitable design could achieve zero visibility capability, without the use of ground based aids.

The US Federal Aviation Administration (FAA) has decided to terminate the development of the microwave landing system (MLS) for US airports, declaring its commitment to a civil GPS service. Instrument landing system (ILS) has been the mainstay for the last forty years. This system no longer meets modern needs, as it is seen to limit airport capacity and efficiency by its requirement for simple straight in approaches. Additionally the potential loss or overlap of ILS frequencies will limit its use. The third generation of communication satellites currently being planned will provide more power and capacity to keep pace with expanding needs and markets. Linked to a suitable ESAS these will do away with the need for any ground based aids.

The Airbus family A330/340 have already being granted European joint airworthiness certification (JAA) for GPS to FAA guidelines. This

GPS system currently provides an accuracy of 100m horizontal position, compared with 600m for conventional VOR/DME. With further development will eventually be reliable enough for bad weather operations to airports with no ILS.

...the hybrid approach!

Flight safety

Aviation has the respect and admiration of the flying public for its extraordinary safety record. It has reached this level of success by identifying and proactively addressing safety issues at all levels of the aviation system. However, with this success comes a tremendous responsibility to maintain and improve this level of safety. Learning from mistakes of others and analyzing accident reports is one way of advancing flight safety. However the first principle of today's precision approach to improving flight safety is that policy decisions should be based on any intelligently gathered and analysed data which takes account of far more than just major accidents and their primary causes. Using flight data recorder (FDA) information, downloaded regularly to gain aircraft health monitoring and operational information, running regular quality control audits, having an effective incident reporting system with communication, feedback and information exchange are all important means to this end.

Once lessons have been extracted from this information priorities can be set for corrective action. Such a structured data-driven and analyzed process should be the objective of all airline management and training departments.

There are still many issues to be addressed. The link between design and training is increasingly important. Compatibility between airborne systems and the air traffic environment needs further scrutiny. Further improvements in multi-crew teamwork and adherence to procedures would certainly yield more safety advantages. Similar aircraft types are operated world-wide by a multitude of different nationalities and diverse line crews' backgrounds. Cultural differences have already been shown to a potential threat to flight safety. Poor training, lack of experience, lack of knowledge, lack of aptitude and poor selection are all part of the cause. These issues, and many others, will continue to exercise the minds of flight deck designers, trainers and regulators for a long time to come.

Postscript

The realization that human error features in the majority of aviation accidents is recognized because the investigation of aircraft accidents is so thorough. A constructive look at the root causes reveals that human lapses predominate, and that controlling these lapses offers the best chance of reducing the accident rate.

Flight safety, now more than ever before, is dependent on company culture, which is the responsibility not only of senior management but also its training staff. If company culture is right, then all employees can operate in an environment that encourages professional integrity and enhances performance.

Computers are playing an ever-increasing role in controlling flight. Technology must serve the pilot and not be his master. There are many recorded instances of the flight computer doing its own thing, with the pilots out of the loop. A computer performing below par, a pilot not up to speed with situational awareness, opens up a whole new range of dangerous possibilities! New technology has relieved the pilot of the many tedious monitoring tasks, but has also created new problems, for which we are not yet fully prepared. There is still no substitute for individual professional competence, sensible selection, sound supervision, good management, and perhaps above all thorough training.

It is hoped that this book will provoke argument and further discussion on this fascinating and remarkable subject.

Bibliography

Abercrombie M.J. *(1989) The Anatomy of Judgement - an investigation into the process of perception and reasoning.* Free Association Books, London

Air Accidents Investigation Branch (AAIB) *(1990) Report of the B737 accident Kegworth 1989.* HMSO

Anderson H.B., Soerenson P.K., Weber S., Soerenson C., *(1966) A study of the perfomance of captains and crews in a full mission simulator.* Roskilde, Denmark, Risoe National Laboratory

Baxter E.P., *(1996) Skill Correction and Accelerated Learning in the Workplace*

Baxter E.P. & Dole S.L., *(1990) Working with the brain – not against it*

Billings C.E., *(1997) Aviation Automation: The search for a human centred approach.* Lawrence Erlbaum Associates

Boeing *(1998) Statistical summary of commercial jet airplane accidents.* Seattle, WA. Boeing Commercial Aeroplane Group

Boeing Aircraft Company *(1990) Flight Study Guides.* Boeing Commercial Aircraft Division, Seattle

Bond N.A., Bryan G.L., Rigney J.W., Warren N.D. *(1985) Aviation Psychology* University of California

British Midland *(1994) Pilot Career Appraisal and Development Scheme CAD,* British Midland, Derby, UK

Civil Aviation Authority *(1998) Global fatal accident review 1980 – 1996.* CAA London

Collins R.L. *(1978) Flying Safely.* Adam and Charles Black, London

Davies D.P. *(1971) Handling the Big Jets.* Air Registration Board

De Bono E. *(1970) Lateral Thinking - A textbook of creativity.* Penguin

Delegation Generale pour l'Armament DGA *(1994) Provisional report on the A330 accident Toulouse August 1994*

Dole S.L., *(1991) New ways for old – teaching mathematics*

Freemantle D. *(1992) Incredible Bosses - the challenge of managing people for incredible results.* McGraw Hill, Maidenhead, UK

Gann E.K. *(1961) Fate is the Hunter.* Hodder and Stoughton

Gibb C.A. *(1969) Leadership.* Penguin

Gibson R.H. *(1989) Managerial Leadership Assessment, Personality Correlates of and Sex Differences in Ratings.* Department of Psychology, University of Texas

Griffin J. *(1996) Becoming an Airline Pilot*

Helmreich R.L., Hines W.E., *(1997) Crew Performance in the Approach and Landing Phase.* Cartegena, Columbia

Helmreich R.L., Merritt A.C., *(1998) Culture at Work in Aviation and Medicine,* Ashgate

Helmreich R.L., Merritt A.C., *(1998) Culture at Work: National, organizational and professional influences.* Ashgate

Helmreich R.L., Wilhelm J.A., *(1989) Status Report on the NASA/UT Project.* Dept. of Psychology, University of Texas at Austin

Helmreich R.L., Wilhelm J.A., *(1989) Validating Personality Constructs for Pilot Selection.* Dept. of Psychology, University of Texas at Austin

Hines W.E., *(1998) Flight Crew Performance in Standard and Automated Aircraft.* University of Texas at Austin

Humphreys G.W., Riddoch M.J., *(1987) To see but not to see - a case study of visual agnosia.* Lawrence Earlbaum Associates, London

International Air Transport Association (IATA) *Aircraft Automation Report: implications of aircraft automation.* IATA July 1994

Jones Nelson R., *(1986) Human Relationship Skills.* Casell, London

Lewis R.D., *(1997) When Cultures Collide.* Nicholas Brealey, London

Lyndon E.H., *(1989) I did it my way.* Australasian Journal of Special Education

Maurino D., Reason J., Johnston., Lee R., *(1995) Beyond Aviation Human Factors.* Ashgate

Maurino D., *(1996) Eighteen Years of CRM Wars. Applied Aviation Psychology. Achievement, change and challenge.* Eds. Hayward B., Lowe A., Sydney A., Avebury Aviation

Merritt A.C., *(1966) National Culture and Work Attitudes in Commercial Aviation.* University of Texas

Merritt A.C., Helmreich R.L., *(1997) CRM: Error, stress, culture.* Seminar: Jakarta, Indonesia

Merritt A.C., Helmreich R.L., Wilhelm J.A., Sherman P.J., *(1996) Flight Management Attitudes Questionnaire.* University of Texas at Austin

Merritt A.C., Helmreich, R.L., *CRM! I hate it! What is it?* Paper presented to the Orient Airlines Association Seminar, Jakarta, April 1996

Morgan J., Welton P., *(1986) See What I Mean - An Introduction to Visual Communication.* Edward Arnold

McDonnell Douglas *(1983) Flight Study Guide.* Mc Donnell Douglas, Commercial Aircraft Division, Long Beach

Murphy E., *(1994) European Association for Aviation Psychology (EAAP).* Conference. Dublin, March 1994. Europilote

NASA *(1997) An analysis of instructor techniques and crew participation -*

facilitating LOS de-briefings – a training manual

National Aeronautics and Space Administration (NASA) *(1989) The Impact of Cockpit Automation on Crew Co-ordination and Communication.* US FAA and NASA

RAeS Human Factors Group *(1988) Guide to Performance Standards for Instructors of Crew Resource Management Training in Commercial Aviation.* Riverprint UK

Reason J., *(1997) Managing the Risks of Organizational Accidents.* Ashgate

Riso D.R., *(1985) The Practical Guide to Personality Types.* Aquarian Press, London

Roger J., McWilliams P., *(1992) Wealth 101 - getting what you want - enjoying what you've got.* Prelude Press, Los Angeles

Rowell J.R., *(1990) Changing misconceptions* International Journal of Science Education

Rudisill M., *(1995) Line pilot's attitudes and experience with flight deck automation.* Columbus, Ohio

Seamster T.L., Boehm-Davis D.A., Holt R.W., Schultz K., *(1998) Developing Advanced Crew Resource Management Training.* FAA

Schutte P.C., Trujillo A.C., *(1996) Flight Crew Task Management*

Sherman P.J., Helmreich R.L., Merritt A.C., *(1997) National culture and flightdeck automation.* International Journal of Aviation Psychology

Sherman P.J., Helmreich R.L., *(1998) Attitudes towards automation - effects of national culture.* University of Texas at Austin

Telfer Ross A., *(1993) Aviation Instruction and Training.* Ashgate

Tullo F., Salmon T., *(1998) The role of the check airman in error management. (Ninth International Symposium on Aviation Psychology)* Columbus, Ohio State University

UK Civil Aviation Authority *(1998) Flight Crew CRM Training Standards (AIC 114/1998)* CAA London

UK Civil Aviation Authority *(1998) Crew Resource Management (AIC 117/1998)* CAA London

UK Civil Aviation Authority *(1990) Notice to AOC Holders No 5 and 11/90.* CAA, London

UK Civil Aviation Authority *(1990) Flight Operations Directive,* 30 May 1990, CAA, London

UK Civil Aviation Authority (1992) *Aeronautical Information Circular (AIC) 80/1992.* CAA, London

Web site links on related subjects:

Web Site: *The CRM Advocate* – published quarterly for the professional air crew
    trainer
        http://www.caar.db.erau.edu/crm/resources/crmadvocate/
Web Site: *RAeS Human Factors Group*
        http://www.raes.org.uk/humanfactors.html
Web Site: *Aeronautical Information Circulars UK*
        http://www.ais.org.uk
Web Site: *Guide to Performance Standards for Instructors of CRM Training in
    Commercial Aviation* – Published jointly by RAeS, CAA & ATA
        http://www.riverprint.co.uk
Web Site: Kray N., Embry-Riddle Aeronautical University. *Industry CRM
    Developers Group*. A forum to identify needs, coordinate processes and
    facilitate development of CRM and Human Factors resources and products.
        http://www.crm-devel.org/avinstr
Web Site: FAA Human Factors Research Management System
        http://hf.faa.gov/
Web Site: Situation awareness & stress tolerance. Discussion group for selection
    of pilots and air traffic controllers.
        http://www.aero.ca/index.html
Web Site: FAA AQP Airline training management system
        http://www.atmssoftware.com/
Web Site: Learning techniques – *Old Way/New Way*
        http://www.personalbest.com.au/~pbaxter/habits.htm
Web Site: University of Texas Crew Reasearch Project. Human Factors in
    aerospace, medicine and other safety-critical work environments.
        http://www.psy.utexas.edu/psy/helmreich/nasaut.htm
Web Site: Bureau of Air Safety Investigation (BASI).Responsible for investigating
    accidents, incidents and safety deficiencies involving civil aircraft. Australasia.
        http://www.basi.gov.au/
Web Site: CAA – consulting and training arm of UK civil aviation authority. Aim
    is to promote the development of safer air transport on a global scale.
        http://www.nats.co.uk/is/fr_content.html
Web Site: Aviation Safety Network – news, comment, accident reports,
    publications.
        http://aviation-safety.net/index.htm
Web Site: Callback. NASA's aviation safety reporting system. (ASRS)
        http://olias.arc.nasa.gov/asrs/callback.html
Web Site: AQP – a turning point in aviation training
        http://www.faa.gov/avr/afs/aqphome.htm
Web Site: Flightdeck automation. Study of HF issues of aircraft flight deck
    automation. Oregan State University
        http://flightdeck.ie.orst.edu/index.html

Web Site: NASA Ames research centre of excellence for information technology.
    http://www.arc.nasa.gov/
Web Site: Understanding how pilots make decisions is the goal of the program
    sponsored by the University of Oregon and NASA-Ames Research Center, and
    funded by the National Science Foundation.
    http://spia.uoregon.edu/idrs/
Web Site: Aspect One. Transportation News & Human Factors Transportation &
    Aviation Industry news are accessible from this site. Human Error, Human
    Factors, Decision-Making, Judgement, and Consequential Litigation issues are
    also explored.
    http://www.wcinet.net/~aspect/
Web Site: Australian Civil Aviation Safety Authority. The Civil Aviation Safety
    Authority (CASA) of Australia maintains this web site as a means of
    disseminating information of interest to the Australian aviation community.
    CASA provides news concerning legislation and regulation, aircraft register,
    corporate aviation, and general aviation in Australia.
    http://www.casa.gov.au/index.htm
Web Site: Aviation Safety Connection. Aviation Safety Connection focuses on
    improved aeronautical decision-making and pilot judgment. This page contains
    a quick link to a data bank containing interactive discussion groups, NTSB
    accident analyses & ASRS incident analyses,
    http://www.aviation.org/
Web Site: Aviation Safety Institute. The Aviation Safety Institute (ASI) is a non-
    profit aviation safety research center headquartered in Worthington, Ohio. Since
    1973, ASI has studied operational incidents which but for one or more missing
    links would be accidents.
    http://www.asionline.org/
Web Site: Aviation Week Air Safety Center. Aviation Week Air Safety Center is
    part of the on-line version of Aviation Week magazine that is published by
    McGraw-Hill. It offers up-to-date news and articles relative to aviation safety.
    SafeLinks offers access to government (DOT, NTSB, FAA, NASA, etc.) and
    non-government (academic, non-profit, and industry) agencies.
    http://www.awgnet.com/safety/
Web Site: Federal Aviation Administration. FAA Office of System Safety. The
    Office of System Safety acts as a coordinator within the FAA on safety issues.
    Its primary function is to help improve aviation safety by facilitating effective
    use of safety data. This site links to aviation safety data, reports, publications,
    and databases.
    http://nasdac.faa.gov/
Web Site: FAA Aviation Safety Information. Links are provided for press
    releases, aviation safety public information, and access to aviation safety
    databases using a powerful search engine.
    http://www.faa.gov/asafety.htm

Web Site: Transport Canada Aviation Safety Services. The Safety Services Branch promotes aviation safety through the delivery of aviation safety education programs. Links to safety brochures and videos.

    http://www.tc.gc.ca/aviation/syssafe/index.htm

Web Site: United States Air Force. Air Force Safety Center.The Air Force Safety Center at Kirtland AFB, NM provides access to military aviation related (flight, ground, and weapons) accident statistics, descriptions of Air Force safety education courses, and provides links to various Air Force safety publications and to the Risk Management Information System (RMIS).

    http://www-afsc.saia.af.mil/

Web Site: United States Navy. Naval Safety Center Operated by Navy and Marine Corps Operational Risk Management (ORM), this site provides information regarding efforts to improve military safety in land, sea, and aviation environments. Links are provided for access to Naval and Marine Corps safety publications, statistical data, and surveys.

    http://www.safetycenter.navy.mil/

Web Site: School of Aviation Safety. The Navy and Marine Corps School of Aviation Safety is a component of the Naval Postgraduate School in Monterey, California. This site provides links to research findings, software, and Navy publications concerning aviation safety.

    http://avsafety.nps.navy.mil/

Web Site: Aircrew Coordination Training. As part of the Naval Aviation Schools Command in Pensacola, Florida, the Navy Aircrew Coordination Training (ACT) and Crew Resource Management (CRM) web page is primarily a source for obtaining class schedules. Behavioral skills such as decision making, assertiveness, communication, mission analysis, adaptability/flexibility, and situational awareness are briefly explored. Links to Crew Resource Management information.

    http://www.act.navy.mil/index.html

Web Site: National Business Aviation Association – Safety. The National Business Aviation Association (NBAA) is a trade association headquartered in Washington, DC that was founded to represent and protect the interests of the business aviation community. The Safety page of NBAA's web site is dedicated to enhancing the safety of business aviation. It facilitates open information exchange regarding safety issues by housing safety reference documents and providing statistics about the safety of business aviation.

    http://www.nbaa.org/safety/

Web Site: Air Line Pilots Association (ALPA). The Air Line Pilots Association is a union representing 53,000 airline pilots at 54 U.S. and Canadian airlines. Site contains news and information related to the airline industry.

    http://www.alpa.org

Web Site: European Collaborative Decision Making Portal. Sponsored by IATA and Eurocontrol, this site provides information on European CDM initiatives and projects and provides links to a variety of European aviation resources.
http://www.euro-cdm.org/

Web Site: Federal Aviation Administration FAA (home page)
http://www.faa.gov

Web Site: FAA Human Factors. This page is maintained by the office of the FAA Chief Scientific and Technical Advisor for Human Factors to provide the aviation community and other interested users with information about human factors research, programs, and applications within government, academia, and industry.
http://www.hf.faa.gov/

Web Site: FAA Civil Aeromedical Institute (CAMI). CAMI is the medical certification, research, and education wing of the FAA. CAMI's mission is to study the factors that influence human performance in the aviation environment, find ways to understand them, and then communicate that understanding to the aviation community.
http://www.cami.jccbi.gov/

Web Site: Flight Safety Foundation. The Flight Safety Foundation is an independent nonprofit organization concerned with aviation safety developments. This site contains related publications, special reports, and a calendar of events.
http://www.flightsafety.org/

Web Site: Institute for Operations Research and the Management Sciences Aviation Applications Section AviatOR Online is the homepage of the Institute for Operations Research and Management Sciences (INFORMS) Section on Aviation Applications. The Aviation Applications Section of INFORMS · encourages the development and dissemination of applications and research in areas relating to aviation.
http://www.agifors.org/informs/

Web Site: International Air Transport Association (IATA) IATA represents the airline industry. Its goals include the promotion of air safety and financial viability for the airline industry.
http://www.iata.org

Web Site: International Civil Aviation Organization (ICAO) ICAO works to develop the principles and techniques of international air navigation to ensure the safe and orderly growth of international air travel.
http://www.icao.org/

Web Site: MITRE Corporation Center for Advanced Aviation System Development. The Center for Advanced Aviation System Development (CAASD)is a Federally Funded Research and Development Center (FFRDC), sponsored by the Federal Aviation Administration.
http://www.caasd.org/index.html

Web Site: National Aeronautics & Space Administration. NASA (home page)
        http://www.nasa.gov
Web Site: NASA Ames Research Center NASA Ames has the agency lead role in
    airspace operations systems, which includes air traffic control and human
    factors.
        http://www.arc.nasa.gov/
Web Site: NASA Aviation Safety Reporting System (ASRS). The ASRS is a
    cooperative program established by the FAA and NASA and administered by
    NASA. Site contains reports on aviation safety issues, the CallBack newsletter,
    and an incident reporting form.
        http://olias.arc.nasa.gov/ASRS/
Web Site: NASA Glenn Research Center. The work of the Center is directed
    toward new propulsion, power, and communications technologies for
    application to aeronautics and space. The Center's major efforts are in subsonic,
    supersonic, hypersonic, general aviation, and high-performance aircraft
    propulsion systems as well as in materials, structures, internal fluid mechanics,
    instrumentation and controls, interdisciplinary technologies, and aircraft icing
    research.
        http://www.grc.nasa.gov/
Web Site: National Transportation Safety Board (NTSB). The NTSB investigates
    every civil aviation accident in the United States and issues safety
    recommendations aimed at preventing future accidents. The NTSB maintains a
    database on civil aviation accidents and conducts studies of transportation safety
    issues of national significance. The NTSB also serves as the "court of appeals"
    for any airman or mechanic whenever certificate action is taken by the FAA.
        http://www.ntsb.gov
Web Site: University Consortium on Atmospheric Research (UCAR). UCAR
    creates, conducts, and coordinates projects to strengthen education, research,
    and
    technology in the atmospheric and related sciences.
        http://www.ucar.edu/

Commercial aviation web sites with extensive aviation links

Landings: http://www.landings.com/aviation.html
AVWeb: http://www.avweb.com/
ProPilot: http://www.propilot.com/

Web Site; book publishing & aviation related subjects

Ashgate: http://www.ashgate.com
Amazon: http://amazon.com
Flight International: http://www.flightinternational.com

Index